Easy Learning

KS3 Science Revision

Levels 5–7

Patricia Miller

About this book

This book has been written to help you prepare for your Key Stage 3 Science Test at the end of Year 9. It contains all the content you need to do well in the two papers.

The book is divided into six Key Idea sections (e.g. Cells), which correspond to The Framework for Teaching Science Years 7, 8 and 9.

Each topic is contained within a double page. There are lots of clear diagrams which will help you understand and remember the content.

The information boxes have been given approximate National Curriculum levels to show you what level the topic is. You still need to know all of the lower level information, even if you are working at a higher level.

Special features

- **Spot check questions** on every double page are a quick way to check that you've taken in the key points. You can find the answers to these on pages 92–93.
- **Top Tips** give extra advice and pick out key facts to help with your revision.
- **Did You Know?** items are there just for fun and a bit of light relief.

Revision and practice

Use this book alongside Collins *Easy Learning KS3 Science Workbook Levels 5–7*. The workbook contains Test-style questions and a practice paper so you can check that you have learnt and understood everything from this revision book.

Published by Collins
An imprint of HarperCollins*Publishers*
77–85 Fulham Palace Road
Hammersmith
London W6 8JB

Browse the complete Collins catalogue at
www.collins.co.uk

10 9 8 7 6 5 4

ISBN-13 978-0-00-723352-6
ISBN-10 0-00-723352-3

British Library Cataloguing in Publication Data
A Catalogue record for this publication is available from the British Library

Written by Patricia Miller
Edited by Mitch Fitton
Design by Graham Brasnett
Illustrations by Kathy Baxendale, Jerry Fowler, David Whittle
Index compiled by Jane Read
Printed and bound in Malaysia by Imago

Acknowledgements
Photographs
The Author and Publishers are grateful to the following for permission to reproduce photographs:

t = top, *b* = bottom, *c* = centre, *l* = left, *r* = right

p.20 *cr*(*t*) Russell D. Curtis/Science Photo Library, *cr*(*c*) Martin Bond/Science Photo Library, *cr*(*b*) Simon Fraser/Science Photo Library. p.63 Still Pictures/Mark Edwards. p.66 *tr* Michael Marten/Science Photo Library; *br* Philip Craven/Robert Harding Picture Library. p.72 *tr*, *cr* Oxford Scientific Films; *br* © George Shelley/CORBIS. p.73 *tr* Carolyn A. Mckeone/Science Photo Library; *br*(*t*), *br*(*l*) Oxford Scientific Films; *br*(*r*) © Duncan Mcewan/naturepl.com. p.74 Oxford Scientific Films. p.75 Christopher Talbot Frank/Oxford Scientific Films. p.80 John Heseltine/Science Photo Library. p.83 Holt Studios. p.89 *cr* Oxford Scientific Films; *br* Sheila Terry/Science Photo Library.

Whilst every effort has been made to trace the copyright holders, in cases where this has been unsuccessful, or if any have inadvertently been overlooked, the Publishers will be pleased to make the necessary arrangements at the first opportunity.

Contents

Cell structure

level 3

Cell structure

There are many different types of cells which all have different jobs. Most cells have certain features in common.

- The **cell membrane** controls what goes in and out of the cell.
- **Cytoplasm** is a kind of chemical soup which makes up most of the cell in both plants and animals.
- The **nucleus** controls the chemical activity of the cell.

level 3

Structure of plant cells

Plant cells have features which are not found in animal cells.

- Plant cells have a **cell wall** as well as a cell membrane.
- The cell wall is made of **cellulose** and helps to keep plants rigid.
- The **vacuole** is a bag-like structure in the centre of the cytoplasm. It holds water and pushes the cytoplasm up against the cell wall to keep the cell rigid.
- **Chloroplasts** contain the green pigment **chlorophyll**. Chlorophyll is used by plants in photosynthesis – the process by which plants make their own food (see page 16).
- Remember! Plants do not have a skeleton – they keep their rigid shape because of the cell wall and the vacuole.

Plant cell

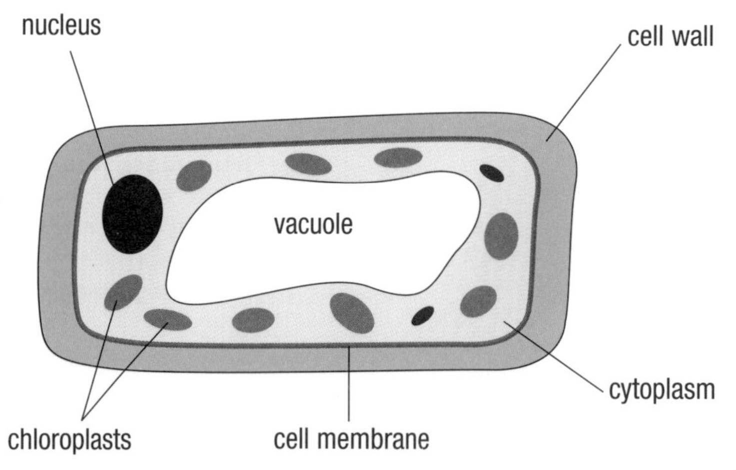

Top Tip!

The basic facts on these two pages are very important and will help you to answer higher level questions on cells.

level 3

Structure of animal cells

- Simple animal cells have the basic structure shown in the diagram below.
- They are different from plant cells because:
  - they have only a cell membrane, not a cell wall
  - they do not have chloroplasts or a vacuole.
- As you will see on page 7, many animal cells have very different structures to enable them to carry out their special jobs.

Top Tip!

You need to know all the level 3 and 4 material in this book to enable you to answer the Test questions fully.

Animal cell

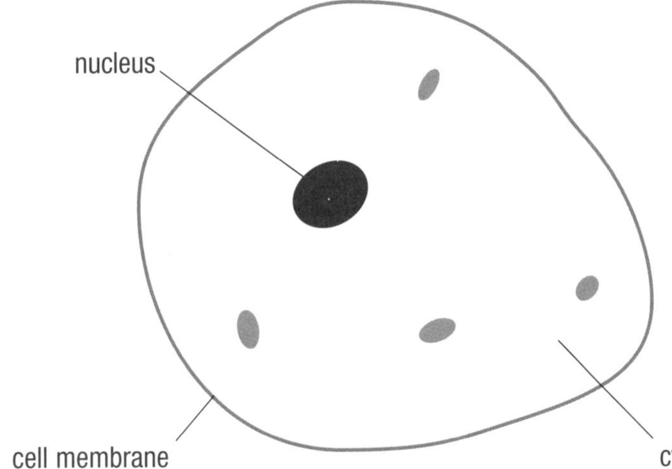

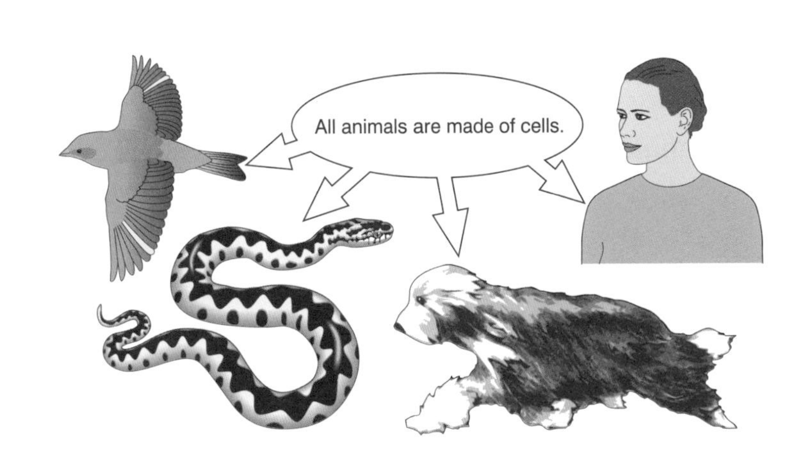

Did You Know?

An amoeba is an animal that is so small it is made from a single cell.

A human body is made of millions of cells.

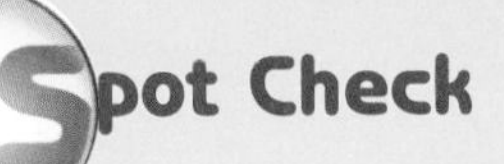

Spot Check

1. Which **three** of these structures are common to animal and plant cells?
   cell membrane **cell wall** **nucleus**
   cytoplasm **chloroplasts** **vacuole**
2. Which part of a plant cell helps the plant to keep its shape even though it does not have a skeleton?
3. Which part of the cell controls what goes in and out of it?

Specialist cells

level 4

Specialist plant cells

Although all plant cells have the same basic structure, many are adapted to carry out a special job within the plant.

levels 4-5

Palisade cells

- **Palisade cells** have a lot of extra chloroplasts to help with **photosynthesis** (see page 16). They are mostly found in the top surface of leaves.

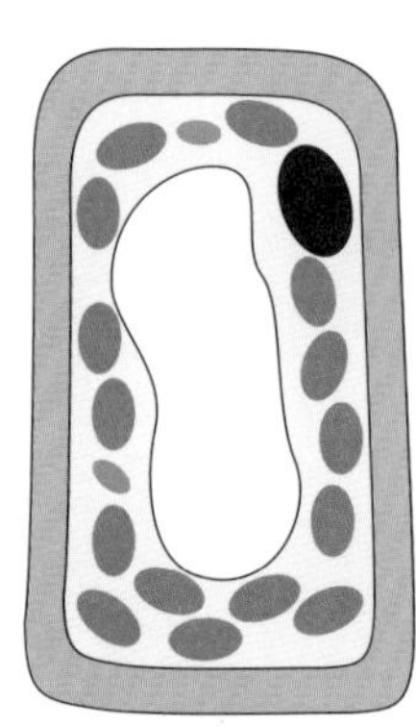

levels 4-5

Root hair cells

- **Root hair cells** have no chloroplasts at all as they do not need chlorophyll to carry out their role. They are long and thin to enable the plant to draw water from deep down in the soil.

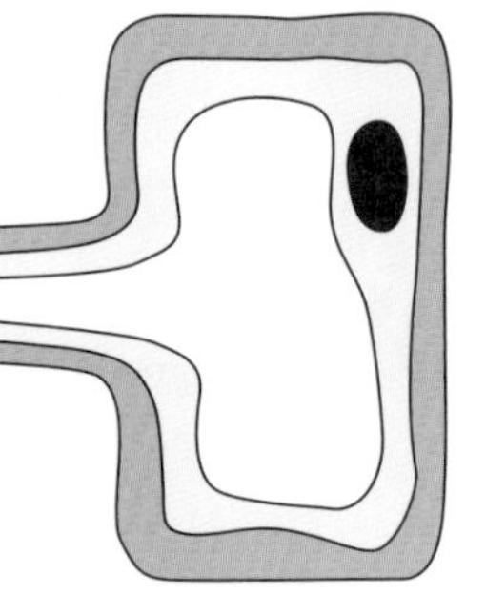

level 4

Size of cells

- On average, single cells are 0.05 mm in diameter and can only be seen under a microscope. However, some cells measure 0.2 mm and certain nerve cells in animals can be as long as 1 m.

level 5

Specialist animal cells

Some animal cells have very complicated adaptations to help them do their job within the animal.

Type of cell	Special task	Adaptation
Nerve cells	Carry messages from the brain to other parts of the body.	Very long and thin – up to 1 metre long.
Ciliated epithelial cells	Keep the airways clean by trapping dust and microbes. Ciliated cells in the oviduct move the egg cell along the tube from ovary to uterus.	Tiny hairs all along the upper surfaces.
Red blood cells	Carry oxygen around in the blood.	No nucleus and bi-concave shape to give the biggest surface area for absorbing oxygen.
Sperm cells	Carry the male genetic information and fertilise the egg to create a new animal.	A tail for swimming and a tough head for penetrating the egg – and half the usual number of chromosomes in the nucleus. Smaller than an egg cell.
Egg cells (ova)	Carry the female genetic information and join with a sperm cell to make a new animal.	Carry a supply of food for energy for the embryo until it is implanted. Half the number of chromosomes in the nucleus just like the sperm cell. When the sperm and the ovum join together in fertilisation, the new cell has the full number of chromosomes.

Did You Know?

When a baby girl is born she has all the eggs she will need already in her body.

Spot Check

1. What is the function of ciliated cells in the lining of the oviduct?
2. Why do egg and sperm cells have only half as much genetic material in the nucleus as other cells?
3. How are red blood cells adapted to carry out their special function?

Tissues and organs

Organisation

level 4

- Apart from the very simplest organisms, plants and animals are made of lots of cells.
- The cells group together to form **tissues**, and tissues group together to form **organs**.

Growth

level 4

- A plant or animal grows because its cells divide and make new cells NOT because the cells get bigger and bigger.

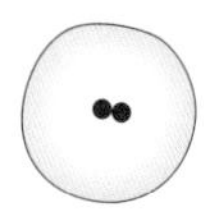

The nucleus of the cells splits in two.

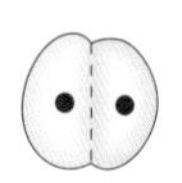

A new cell membrane forms in the middle.

The new daughter cells get bigger.

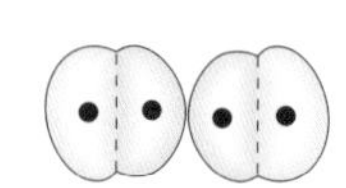

Once the daughter cells are full-size, they too can divide.

Tissues

level 4

- Cells have different structures so that they can perform different functions in the animal or plant.
- Groups of the same cell join together to make tissues.

Examples of tissues

level 4

- Muscle cells are grouped to form muscle tissue.

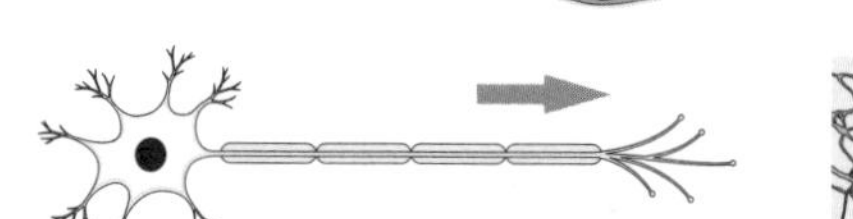

- Nerve cells join together to make the nerve tissues which carry messages from the brain to all parts of the body.

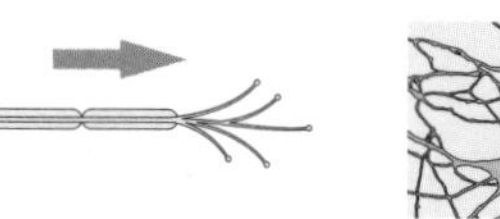

- In plants, palisade cells (see page 6) cluster near the top of the leaf to enable the plant to make food.
- Root hair cells join together to make root hair tissue.

Top Tip!

One organ may contain more than one type of tissue. The heart, for example, contains muscle tissue and nerve tissue.

level 4

Organs

- Groups of different tissues join together to form organs.
- The heart, lungs, liver, brain and stomach are all examples of organs. They each have a particular job to do and consist of groups of tissues made of specialist cells to enable them to do that job.
- In plants, the leaves and roots are organs.

levels 4-5

Systems

- Everything that living things do relies on an organ or a group of organs. A group of organs working together makes up a **system**. In mammals it works like this:

Characteristics of living things	Systems that enable this to happen	Organs in this system
Movement	Skeletal system in animals	Bones and muscles
Respiration	Respiratory system	Lungs
Sensitivity	Nervous system	Brain, nerves and the sensing organs including eyes, ears and the skin.
Growth	All systems – growth happens by cell division.	All the specialist tissues that make up the organs of the body.
Reproduction	Reproductive system	Reproductive organs such as the ovaries and womb in females and the penis and testes in males.
Excretion	Excretory system	The kidneys are the most obvious organs but the lungs and the skin are also used by the body to get rid of waste.
Nutrition	Digestive system allows food to be broken down so that the body can make use of it.	The organs involved in digestion include the mouth, stomach and intestines.

Did You Know?

Dolphins can hear sounds made by other dolphins and whales over distances up to 300 miles.

1. Which part of a cell divides first when a cell divides to enable an organism to grow?
2. Which organs make up the nervous system?
3. Name **three** organs that allow humans to excrete waste products.

The digestive system

level 3

Food

- Plants make their own food using the energy from sunlight (see page 16) but animals have to take in food and break it down to release energy.

level 4

Digestion

- **Digestion** starts in the mouth where food is broken down into smaller pieces. Enzymes in **saliva** chemically break down starch.
- From the mouth food travels down the **oesophagus** until it reaches the **stomach**.

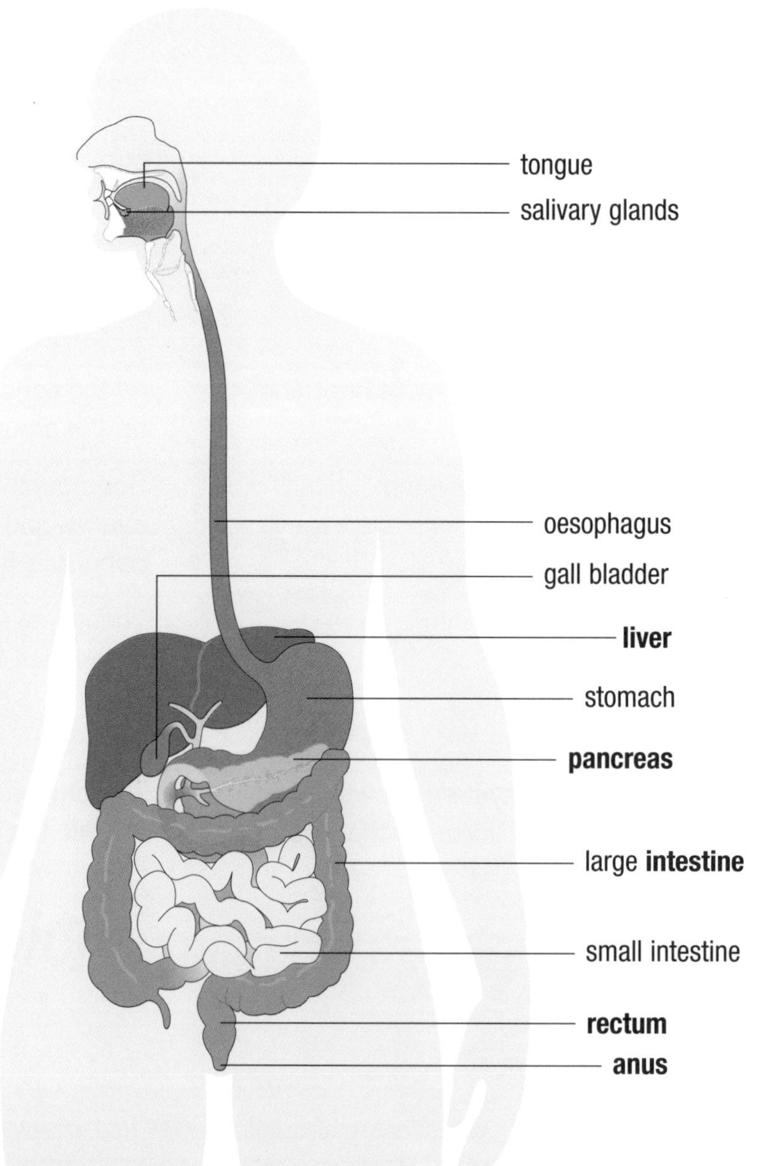

levels 4-5

Organs of the digestive system

Organ	Function in digestion
Digestion starts when food enters the mouth. The food then moves down the oesophagus to …	
Stomach	Muscular walls churn up the food. At the same time gastric juices mix with the chopped up food to produce a sort of mushy, soup-like substance which then passes into the small intestine. Protein digestion starts in the stomach.
Liver	The food does not pass through the liver on its way through the body – but the liver makes a fluid called bile which helps in the breakdown of fats. Bile is stored in the gall bladder.
Small intestine	This is where the mush from the stomach is further broken down so that the molecules become small enough to pass through the intestine wall into the bloodstream. Enzymes from the pancreas help with this chemical breakdown.
Large intestine	In this final stage, water is absorbed back into the body and the remaining solid waste is passed into the rectum where it is stored until it is passed out of the body through the anus.

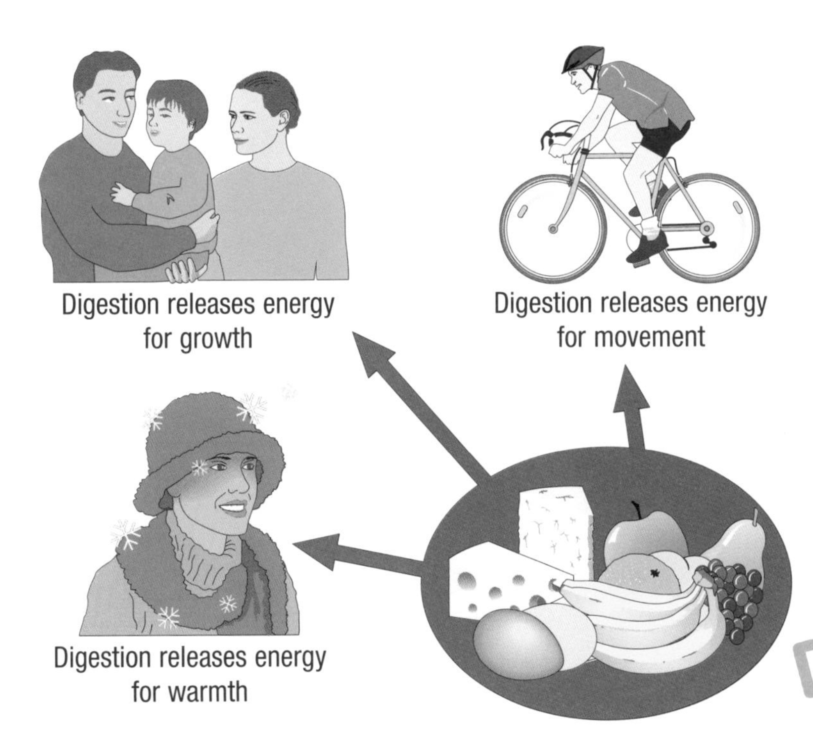

Top Tip!

Anything to do with the stomach is described as 'gastric'.
The chemicals that break down the molecules in food into smaller molecules are called enzymes.

Did You Know?

If the tubes in a French Horn were straightened out it would be over 6 m long – and human intestines are a little bit longer.

Spot Check

1. What part does the liver play in digestion?
2. In which organ of the digestive system is water absorbed back into the body?
3. What is the scientific word used to describe anything to do with the stomach?

CELLS The reproductive system

level 4

Organs of reproduction in humans

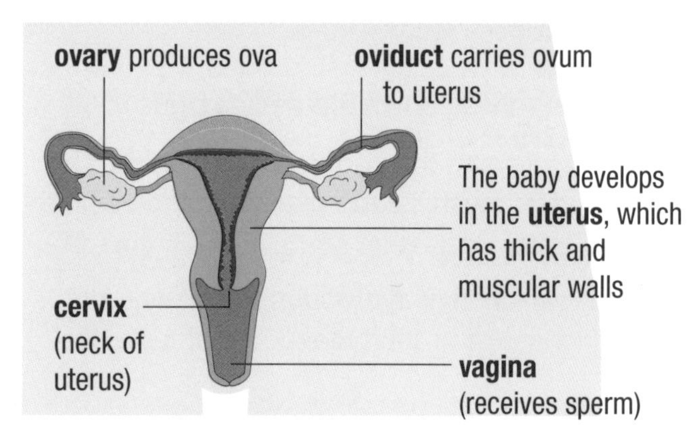

Female reproductive organs

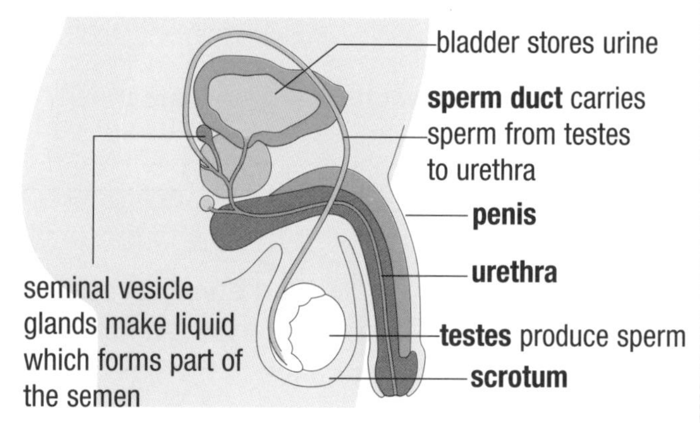

Male reproductive organs

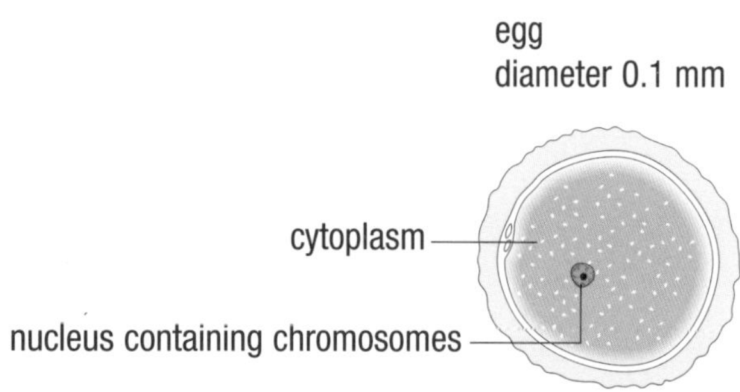

A human foetus takes about 9 months to develop in its mother's uterus – for elephants this gestation period is 22 months, but for hamsters it is only 16 days!

levels 4-5

Fertilisation

- A new human being is made when an egg cell (or ovum) in a female is fused with a sperm cell from a male. This fusion is called **fertilisation**.
- The chromosomes contain the DNA – the genetic material. The new human gets half its DNA from each parent.

1. How long does a human pregnancy last?
2. Why is it dangerous for a pregnant woman to smoke cigarettes?
3. What happens at around day 14 of the menstrual cycle?

levels 5-6

The menstrual cycle

- An adult female releases an ovum from one of her ovaries about every 28 days.
- Before the ovum is released, the uterus prepares to accept a fertilised ovum – just in case fertilisation takes place. These events are part of the **menstrual** cycle.

Time in cycle	What happens
Day 0–5	If there is no fertilised ovum, the uterus lining breaks down and is passed out of the vagina. This is called a period or menstruation.
Day 6–13	Uterus lining starts to thicken in case it has to receive a fertilised ovum.
Day 14	Ovulation – a mature ovum is released from the ovary.
Day 15	The ovum starts to move along the oviduct.
Day 16–21	The uterus lining remains thick with a good blood supply in case a fertilised ovum needs to be implanted.
Day 22–28	The lining of the uterus stops developing and if there is no fertilised ovum it starts to break down and the cycle begins again.

level 6

Pregnancy and birth

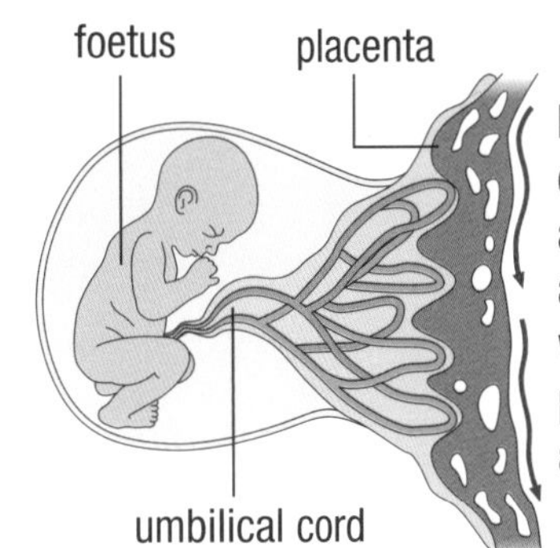

- If an egg is fertilised in the oviduct it starts to divide straight away. By the time it reaches the uterus it is a ball of cells ready to implant in the prepared lining of the uterus.
- After about 6 weeks, the tiny **embryo** floats in a sac of amniotic fluid that protects it until it is ready to be born. The developing **foetus** gets the nutrients and oxygen it needs from the **placenta** via the **umbilical cord**.
- Approximately 9 months (40 weeks) after fertilisation the baby is ready to be born. It is pushed out through the vagina by strong muscular contractions of the wall of the uterus.
- If a pregnant woman drinks, smokes or takes drugs, these substances will pass through her placenta to the developing baby and can affect the baby's birth weight. The baby is more likely to be ill in early life and in extreme cases will be born already addicted to whatever its mother has been taking.

Top Tip!

At level 6 you will need to be able to explain the link between the organs of the reproductive system and things such as the effect that drugs and alcohol can have on a developing foetus.

The circulation system

levels 4-5

Organs of the circulation system

- The major organ in the circulation system is the **heart**.
- The heart is a powerful muscular pump which sends blood to all of the body.
- 'Pulmonary' means to do with the lungs. The pulmonary artery carries blood to the lungs and the pulmonary vein carries blood away from the lungs.

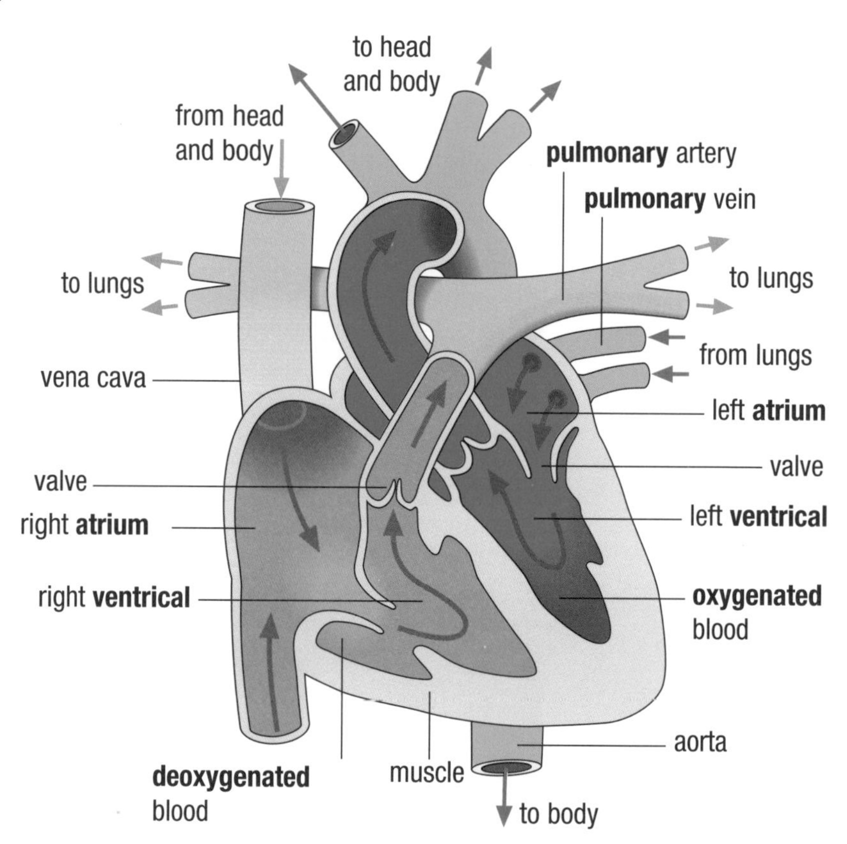

Diagrams of the heart are always drawn as if you are looking at a heart inside a person's chest. This means the left side of the heart is always drawn on the right-hand side of the page.

levels 5-6

Gas exchange

- The blood goes from the right-hand side of the heart to the lungs. Here the red blood cells collect oxygen using a chemical called **haemoglobin**. This **oxygenated blood** goes through the left-hand side of the heart and is sent all around the body.
- Each cell in the body exchanges oxygen for carbon dioxide. The **deoxygenated blood** then comes back to the right-hand side of the heart. From there it is sent back to the lungs where it exchanges its carbon dioxide for oxygen – and the cycle begins again.

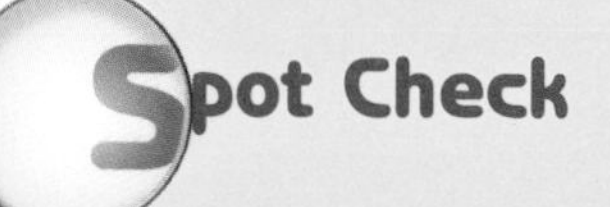

1. From which side of the heart does blood go to the lungs and why?
2. How thick are the walls of capillaries – and why?
3. One of the harmful substances in cigarette smoke is carbon monoxide, which can combine with the haemoglobin in red blood cells. Can you suggest why this is harmful?

level 5

Blood vessel functions

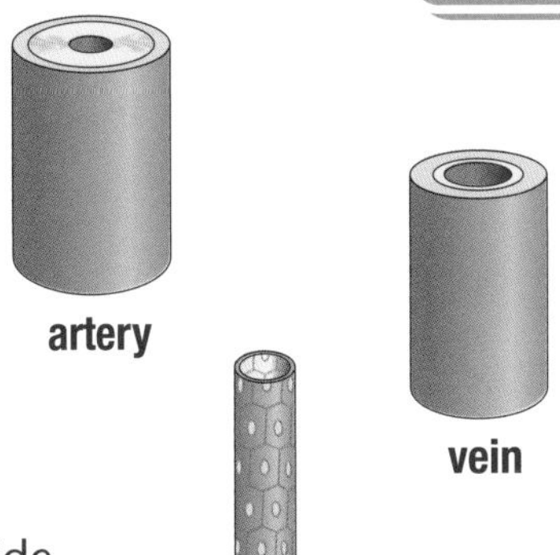

- Thick muscular walls allow the **arteries** to carry blood at high pressure as it is pumped away from the heart.
- **Veins** have thinner walls as the blood they carry is moving more slowly and at a lower pressure going back towards the heart. Veins have valves inside their walls so the blood does not flow in the wrong direction.
- **Capillaries** have walls only one cell thick so that oxygen, carbon dioxide and other food and waste products can pass in and out of them easily.

levels 5-7

Respiration and breathing

- **Respiration** and breathing are NOT the same thing although they are connected – so it is an easy mistake to make.
- Breathing means taking air, which contains about 20% oxygen, into our lungs and expelling air, with little oxygen and a lot of carbon dioxide in it, from our lungs.
- Respiration is the chemical reaction which takes place in every cell in our bodies to release energy from food. We need oxygen for this reaction and carbon dioxide is given out by the cells as a result of this reaction.

Did You Know?

The surface area of the lungs of an adult human is almost 100 m^2 – and the same person's skin will be only about 2 m^2.

level 6

Systems working together

- In the lungs, oxygen is transferred from the air we breathe into the bloodstream to be carried around the body for respiration. Carbon dioxide is carried back to the lungs in the bloodstream from where it is passed to the lungs and breathed out of the body.
- The lungs are part of the respiratory or breathing system – they work with the circulation system to deliver oxygen to the cells for respiration. All the systems work together to form a whole **organism**, made up from millions of tiny cells, all working together.

CELL → TISSUE → ORGAN → SYSTEM → ORGANISM

- The organs in the circulation system work together to respond to different situations. For example, when a person exercises, the heart beats faster so that blood carrying oxygen and glucose reaches the muscle cells more quickly.
- Harmful substances taken in through our lungs, such as those in cigarette smoke, can both damage the lungs and make this process less efficient, allowing poisons to get into the blood and so to every cell in the body.

Photosynthesis

level 3

Sunlight

Photosynthesis is the process by which green plants make their own food. They make use of our principal energy source – the Sun.

- Green plants are able to make their own food. This is why all food chains begin with green plants – called the **producers**.

levels 3-4

Chlorophyll

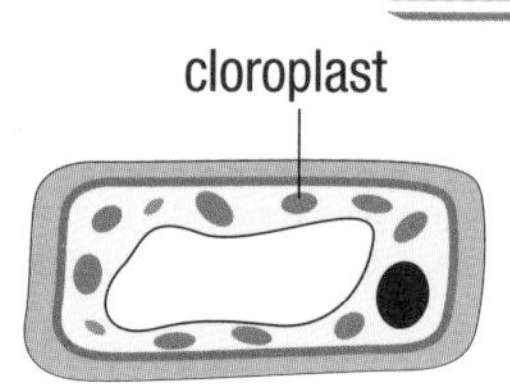

- Plants make food in the cells that contain **chloroplasts**. Chloroplasts are green because they contain the green pigment **chlorophyll**.
- Cells near the top of the leaf always have a large number of chloroplasts so that they can absorb a lot of sunlight for photosynthesis.
- Root hair cells have no chloroplasts – being far away from the sunlight they do not take part in photosynthesis.
- Plants that have **variegated** leaves – that is leaves that are part green and part white – only carry out photosynthesis in the green part of these leaves.

variegated leaf

level 4

Water and carbon dioxide

- As well as chlorophyll and sunlight, a plant needs two important ingredients for photosynthesis: water and carbon dioxide. As it is a chemical reaction, we call the ingredients 'reactants'.

level 4

Plants and food

- Plants make their own food through photosynthesis. They do NOT take in food through their roots. Garden centres sell containers labelled 'Plant Food'. These contain the minerals and trace elements that plants need to stay healthy, BUT plant 'food' is the glucose they make by photosynthesis.

Top Tip!

All green plants carry out photosynthesis and the chemical reaction is always the same. Don't be put off by unfamiliar plant names in Test questions!

level 5

Photosynthesis as a chemical reaction

- Photosynthesis mostly happens in leaves, although any plant cell that has chloroplasts can carry out photosynthesis.
- We can describe photosynthesis using a word equation like this:

Water + carbon dioxide $\xrightarrow[\textbf{chlorophyll}]{\textbf{light energy}}$ **glucose + oxygen**

- This means that the cells of a green plant take in water and carbon dioxide. Providing there is light and chlorophyll present they can make glucose. Oxygen is released as a waste product.

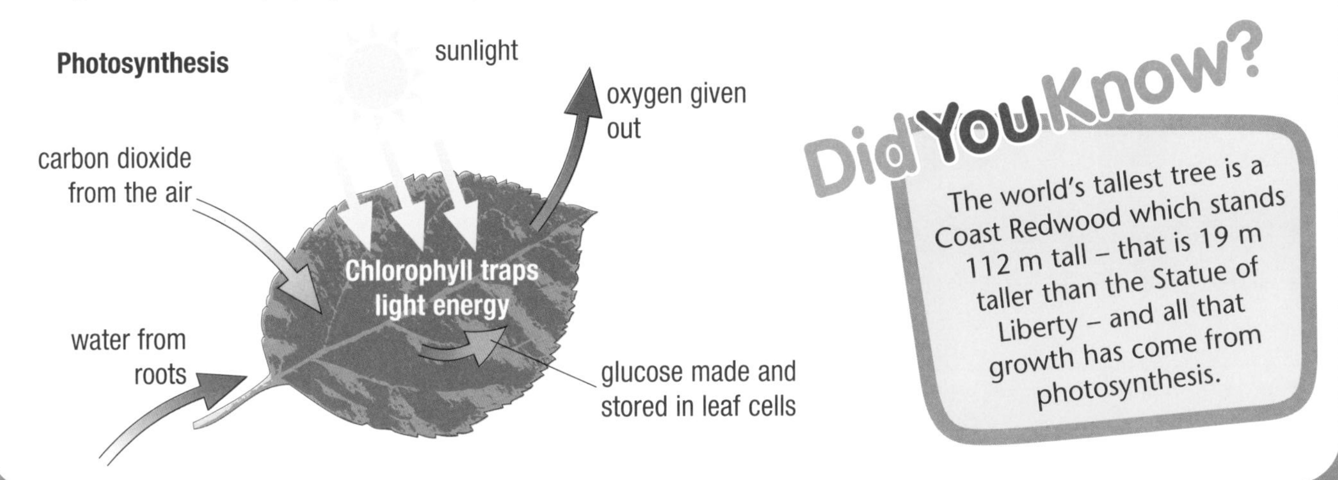

The world's tallest tree is a Coast Redwood which stands 112 m tall – that is 19 m taller than the Statue of Liberty – and all that growth has come from photosynthesis.

level 6

Photosynthesis and respiration

- Plants need to take in carbon dioxide for photosynthesis, and they give out oxygen. Plant cells also carry out respiration to free the energy from glucose. For this they need to take in oxygen and give out carbon dioxide. The equation for respiration is:

  Glucose + oxygen → carbon dioxide + water AND ENERGY IS RELEASED

- If you look carefully at the two equations you will see that they are very much the same just the opposite way around. It is important to remember these two equations and how they are connected.
- During the day, there will be more photosynthesis than respiration. The plant will therefore take in more carbon dioxide than it gives out, and will give out more oxygen than it takes in. At night, there is only respiration taking place, so plants only take in oxygen and give out carbon dioxide.
- Photosynthesis is one of the most important chemical reactions on earth. It is how we trap the energy from sunlight AND how we replace the oxygen we use up in respiration.

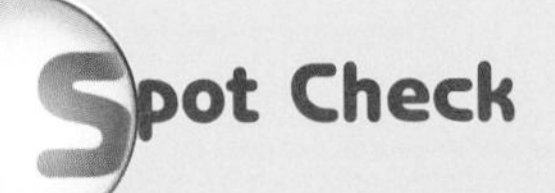

1. If a plant has variegated leaves, what does this mean?
2. What is the word equation for photosynthesis?
3. Give **two** reasons why photosynthesis is such an important chemical reaction.

Photosynthesis and respiration as chemical reactions

level 7

Measuring the rate of the reactions

- One way of measuring the rate of photosynthesis is to measure the rate at which oxygen is produced.
- The rate of respiration can be calculated from carbon dioxide production.
- To find out which reactions are occurring and at what rate in the test tubes below, you need to predict which gas is given off in each tube and how this might change as the conditions change.

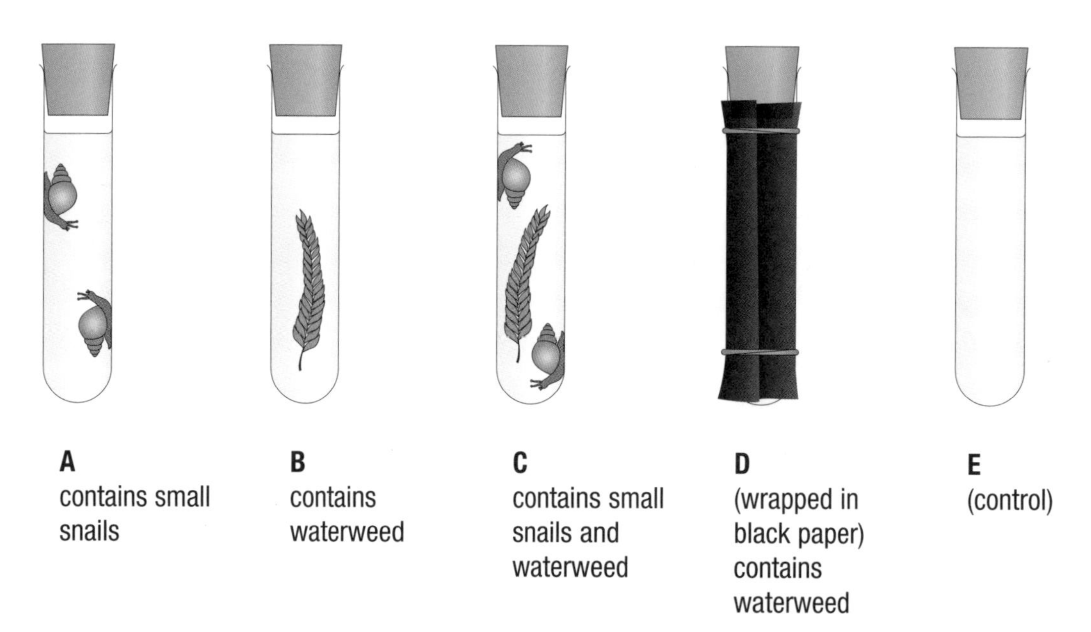

- In tube **A** there would be an increase in carbon dioxide – respiration is taking place.
- In tube **B** there would be an increase in oxygen and a decrease in carbon dioxide – photosynthesis is taking place.
- In tube **C** the levels of both gases would probably be the same – photosynthesis and respiration are taking place at about the same rate.
- In tube **D** there would be an increase in carbon dioxide as the waterweed would be respiring but would not carry out photosynthesis as there is no light.
- In tube **E** the levels of the two gases would not change – as neither photosynthesis nor respiration are occurring.

Top Tip!

Plants respire all the time, but they only photosynthesise during the day. For plants to grow, they must make more food during photosynthesis than they break down in respiration.

level 7

Photosynthesis and competition

All green plants need light to carry out photosynthesis to make food.

- This affects the ways in which different plants live together in the same habitat. Think of a woodland in spring – the trees are still quite bare but the days are becoming longer and sunnier. At this time, flowers such as bluebells and primroses grow and flower. Later, as the leaves on the trees grow and block out the sunlight, these flowers become dormant again.

level 7

Photosynthesis and light intensity

- Some plants have adapted to grow better in conditions of lower light intensity than others. It is important that you can use your knowledge of photosynthesis to explain the growth patterns of plants. The higher the rate of photosynthesis, the faster the rate of plant growth.
- For example, buttercup plants grow mainly in open fields. Dog's mercury is a plant which grows mainly in woodland. The graph shows how the rate of photosynthesis in these two plants changes as the light intensity changes.
- The graph shows how the buttercup grows best at a high level of light intensity while the dog's mercury reaches its highest rate of photosynthesis – and therefore of growth – at much lower light intensity levels. This is why buttercups grow best in open fields in bright sunlight and dog's mercury grows best in woods where the light intensity is lower because of the shade cast by leaves on the trees.

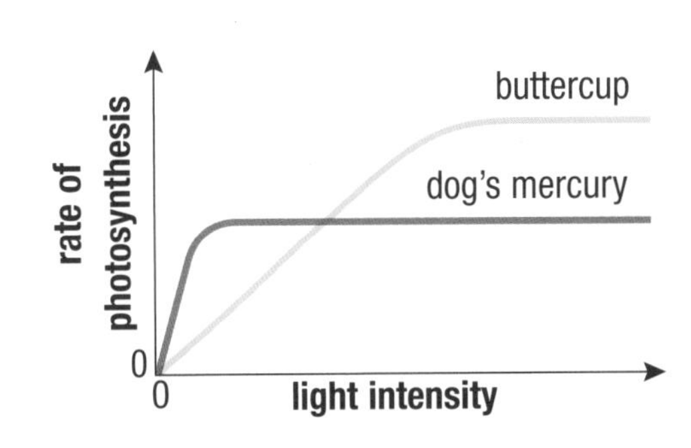

Did You Know?

The Sun is the ultimate source of all our energy here on Earth. Sunlight takes just over 8 minutes to reach the Earth from the Sun.

1. What element is a reactant in respiration and a product of photosynthesis?
2. Why in the past were plants and flowers taken out of the rooms of ill people at night?

Energy resources

levels 5-6

Non-renewable energy resources

- These are also known as **fossil fuels**: **coal**, **oil** and **natural gas**.
- The advantages of using these fuels are:
  - that they contain a lot of stored chemical energy
  - they can be transported and used anywhere.
- The disadvantage of using these fuels are:
  - that they produce waste gases which are harmful to the environment
  - because they took millions of years to produce it is not possible to make any more and supplies will one day run out.

Top Tip!

Don't confuse sources of energy with energy itself. Fuels are not energy, they are sources of energy.

levels 5-6

Renewable energy resources

- Renewable energy resources are:
  - **solar** power
  - **wind** power
  - **tidal** and **wave** power
  - water stored behind dams for **hydroelectric** power
  - **geothermal** power (energy from hot rocks)
  - **biomass** (wood, straw, etc.)
- The advantages of using these energy sources are:
  - that they do not produce the harmful gases that result from burning fossil fuels
  - that they are in no danger of running out.
- The disadvantages are:
  - that they are not such concentrated sources of energy
  - generally they can only be used in places where the conditions are right.

Solar power

Geothermal power

level 6

Renewable energy problems

- Renewable energy is not so flexible as non-renewable energy:
  - solar panels can only be used in the sunshine
  - there are only two places in Britain where it is possible to make use of tidal power.
- Although they do not produce harmful gases, there can still be damage to the environment:
  - some people think wind turbines are ugly so they don't want to live near them
  - using tidal power might mean flooding areas used by sea birds as nesting grounds.

level 5

Energy resources and the environment

- When fossil fuels are burned to release their store of energy, a large amount of carbon dioxide is released into the air. This carbon dioxide goes up into the atmosphere and forms an insulating layer.
- This means that, although the heat energy from the Sun can get into the Earth's atmosphere, the lower energy (reflected waves of radiant heat) cannot get out again. (See also Energy transfer on page 28.)
- This is called the **greenhouse effect** and the increase of this effect is the cause of **global warming**.
- Burning fossil fuels also releases sulphur and sulphur dioxide. This can combine with water in the clouds to produce **acid rain** which damages trees and buildings.
- Renewable energy resources are mostly much cleaner and do not have any of these effects on the environment.
- However, burning biomass such as wood and straw does release carbon dioxide.

carbon dioxide layer reduces the heat energy that is radiated back into space

heat energy from Earth radiates into space

some energy is reflected back to the Earth's surface

heat energy from the Sun

Nuclear power

- The energy stored in the atoms of radioactive materials, such as uranium (a naturally occurring metallic element), can be released to produce energy to generate electricity.
- **Nuclear** power is not really renewable – but it is not a fossil fuel either.

Did You Know?

On Earth we get 700 000 000 000 000 000 000 (seven hundred million million million) watts of energy from the Sun every second.

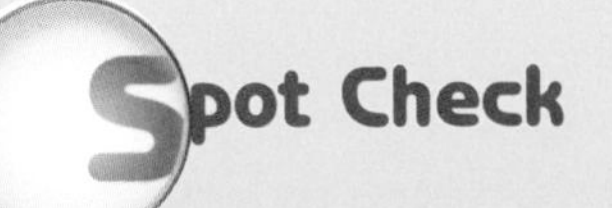

1. Name **one** advantage and **one** disadvantage of using fossil fuels.
2. Which disadvantage of fossil fuels is also a problem when biomass is burned as a fuel?
3. What name is given to the effect of the layer of CO_2 that forms in the Earth's atmosphere?

Generating and using electricity

levels 4-5

Using fossil fuels to generate electricity

One of the most important uses of fossil fuels and renewable energy resources is to **generate** electricity.

- Coal, oil and gas are all used in power stations to generate electricity. The energy stored in coal, oil and gas is chemical energy.
- Whichever fuel is used, the process is always the same. Burning fossil fuels is the most efficient way of generating electricity – but burning fossil fuels causes environmental damage (see page 21).

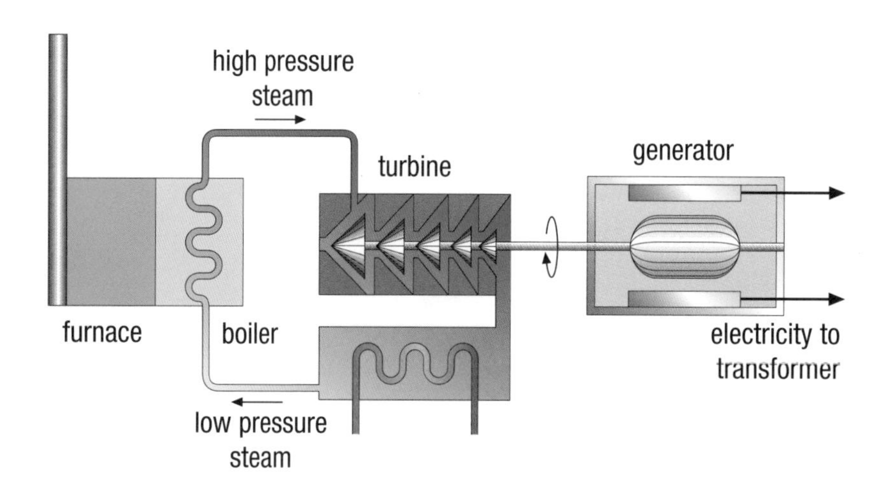

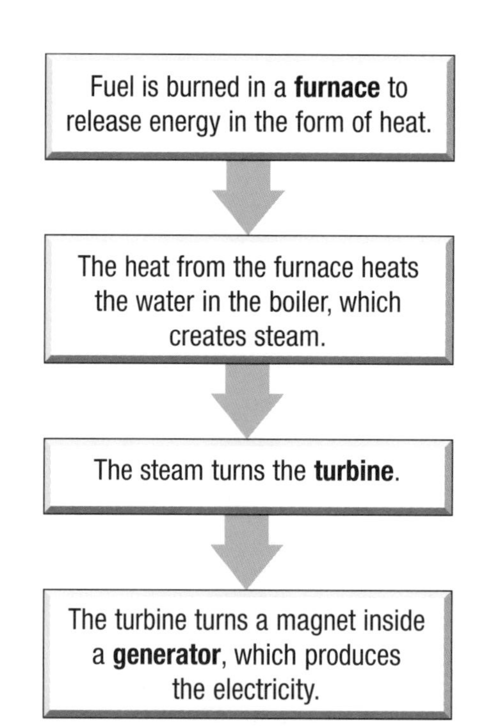

Top Tip!

Remember that electricity is a kind of energy that has to be produced using another energy source.

levels 4-5

Generating electricity from renewable energy resources

- Using wind turbines, tidal or wave power or water stored behind a dam to generate **hydroelectric** power relies on turning the turbine directly. There is no burning of fuel to heat water to make the steam to turn the turbine.
- **Solar panels** trap the energy of the Sun and use that to make electricity.
- These are more environmentally friendly ways of generating electricity but they are much less efficient.

Did You Know?

If 1 million homes all had solar panels it would still only produce 0.1% of the 50 thousand million watts of electrical energy we use in the UK each year.

levels 4-5

Using electricity

- We use electricity for all sorts of purposes at home, at school and in the workplace.
- To conserve fossil fuel supplies and to reduce environmental damage, we need to try to cut down on the amount of electricity we use.
- These are examples of ways to reduce the energy we use:

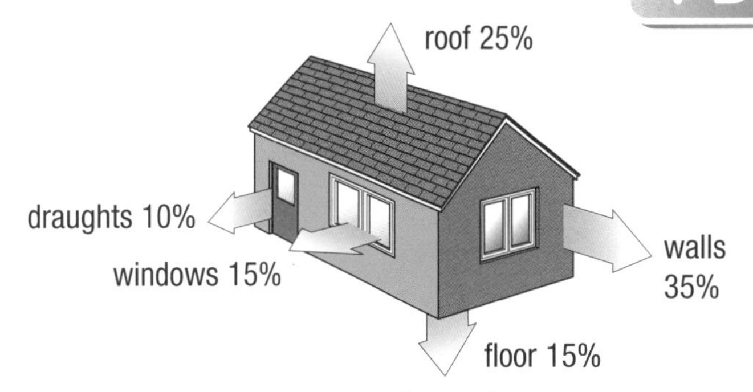

Heat loss from a house

Example of energy use	Way of reducing energy use
Heating our homes	Good insulation reduces the heat that is wasted. This can be done through double glazing, loft insulation and remembering to shut the doors and windows.
Lighting our homes	Use energy efficient light bulbs and turn off lights that are not necessary.
Washing ourselves and our clothes	Baths use much more hot water than showers. Always put a full load in the washing machine.
Boiling the kettle	Only boil the amount you need – don't boil a full kettle to make one cup of tea.

1. What does steam do in a power station?
2. What is the biggest difference in the process of generating electricity from renewable energy resources such as wind, tidal or wave power rather than from using fossil fuels?
3. Name **three** ways of reducing heat loss from our homes.

Electrical circuits

level 3

Simple electric circuits

In order to work, simple electric **circuits** must be complete. Circuits must be made of material that can carry electricity, which are called **conductors**.

You need to know these symbols to draw and understand circuits.

Symbol	Component
+ −	A single cell
	A battery of cells
	A lamp or bulb
	A connecting wire
	A switch
A	An ammeter

As electrical circuits are so widely used, you may be asked a question about circuits in an unfamiliar context – for example, Christmas tree lights or a rear windscreen heater in a car. So read the question carefully and don't be put off.

level 5

Series and parallel circuits

- In a circuit, all the components can be one after the other – a **series** circuit. Or the circuit can have two or more branches with components in each of the branches – a **parallel** circuit.
- The letter I on a circuit diagram means the current.

  On this diagram $I_1 = I_2 = I_3$ as the current is the same at all points in the circuit. So the reading on all the ammeters will be the same.
- In a series circuit if one bulb breaks, it breaks the whole circuit and all the bulbs go out.
- In a parallel circuit, the current divides into two branches, so $I_1 = I_2 + I_3$. If one of the bulbs breaks, there will still be a complete circuit to the other bulb, so it will stay alight.

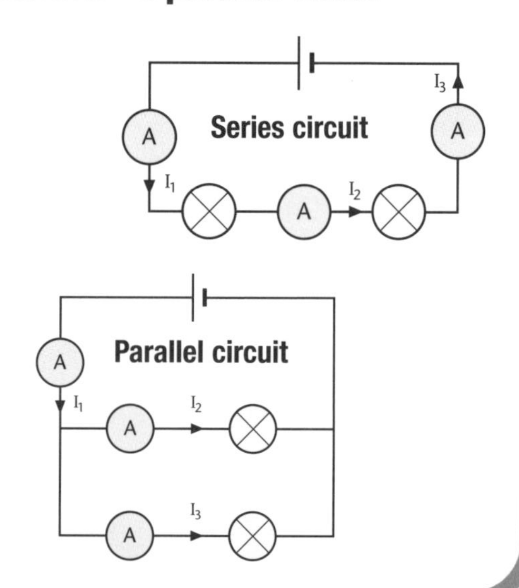

1. Which sort of circuit would you choose if you wanted to make a string of fairy lights – and why?
2. If you went into a room and there were two lights which were both switched off, how could you tell which one had been off all the time and which one had just been switched off?
3. What sort of energy is stored in a battery?

Did You Know?

Bee keepers never put their hives near to electric power lines as this can affect the ability of the bees to make their honey!

level 7

Circuits and current

- Two bulbs in a series circuit will have half the current when compared with a circuit with the same number of cells/battery and only one bulb. Each of the two bulbs will be much less bright than one bulb will be on its own in that circuit.
- Two bulbs in parallel will each be as bright as a single bulb in a series circuit.

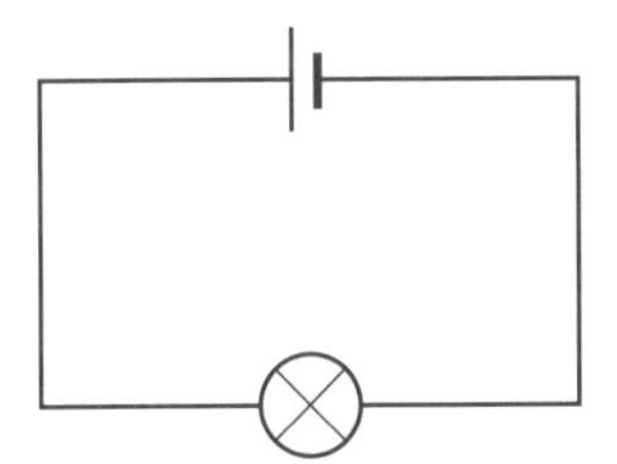

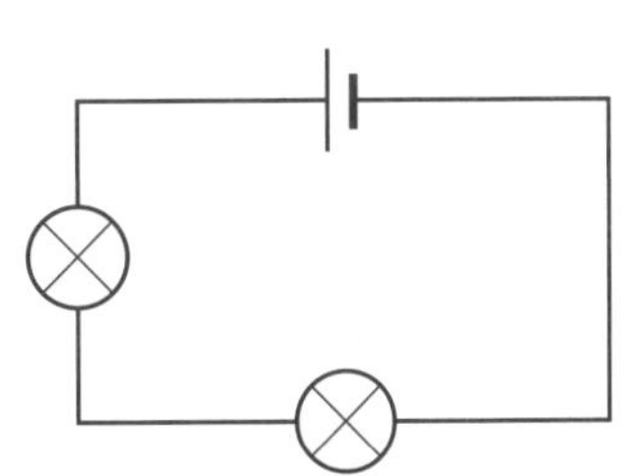

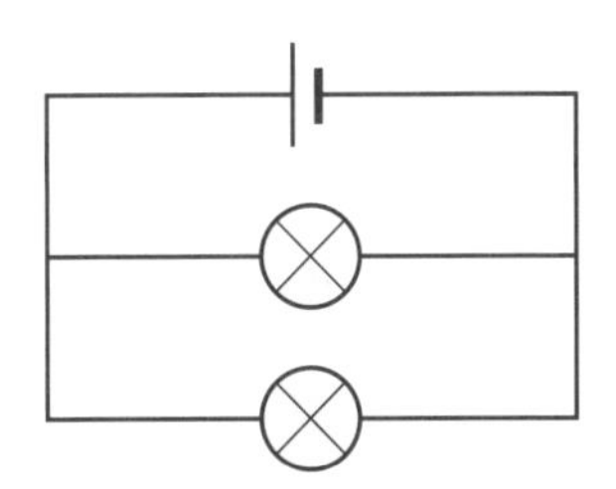

levels 4-5

Energy transfers in electrical circuits

- An electric circuit is a means of transferring energy. As with all energy transfers, some energy is always lost to the surroundings as heat. All electrical appliances get hot, even if heat is not the type of energy that we want.
- A **battery** is a store of chemical energy. When we say a battery has used up all its energy, we do not mean that the energy has been destroyed – it can't be. We mean that the store of chemical energy in the battery has all been transferred by the circuit to the other components and has provided

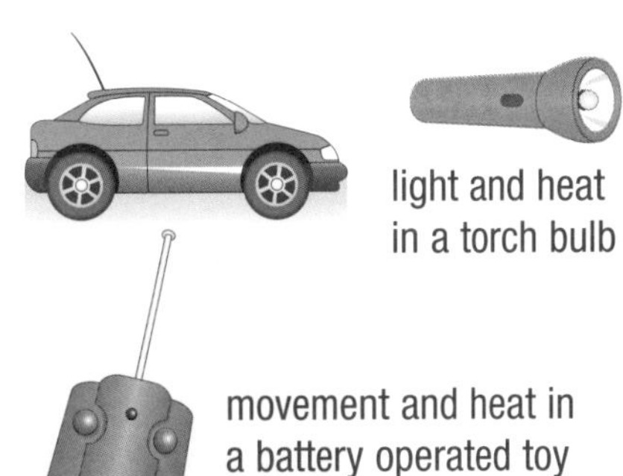

sound and heat in a radio

level 7

Dynamos

- We can give electrical energy to a circuit in other ways. A **dynamo** on a bicycle transfers kinetic or movement energy to electrical energy. It uses the friction of the tyres to turn a magnet inside a coil of wire and make an electric current – just as in a power station, only much much smaller. The supply of energy cannot run out – unlike a battery – but if you stop pedalling the lights go out!

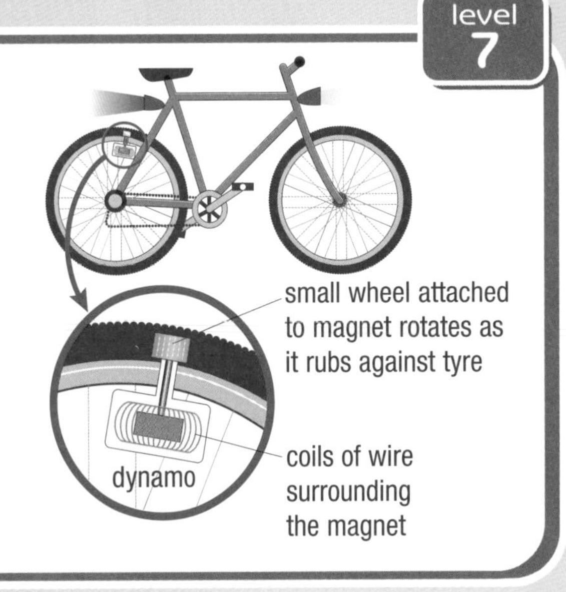

Energy transformation

level 5

Energy conservation

- When energy moves from place to place it sometimes changes its form, but it is never created or destroyed. You will always have the same total amount of energy at the end of a series of transfers as you had at the beginning, even if it is not always in the form that you want.
- Energy can exist in a number of different forms. Some of them are listed below and on page 27.

level 5

Thermal energy

- Heat energy or **thermal** energy is always there during every energy transfer – some of the energy is transformed into heat energy and given out to the surroundings. This is often seen as lost energy as there is not always enough to be measurable but it is still there. Thermal energy is transferred in a number of different ways (see page 29).

level 5

Light and sound energy

- Light energy or visible light is a kind of electromagnetic radiation which travels as a wave and transfers energy from place to place (see page 33).
- Sound energy is also a wave that transfers energy from place to place. Sound waves need particles to vibrate in order for them to travel – unlike light (see page 32).

Light energy

sunlight
sunlight used to evaporate water from leaves
sunlight reflected off leaf
energy absorbed and used in photosynthesis
energy transmitted through leaf
energy lost in respiration as heat

Sound energy

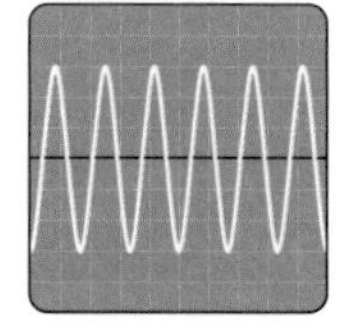

loud sound high frequency

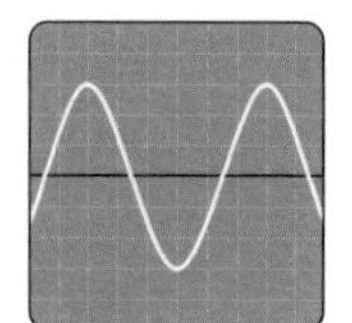

loud sound low frequency

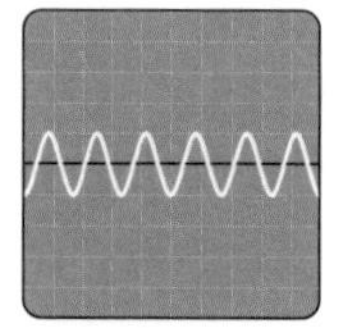

quiet sound high frequency

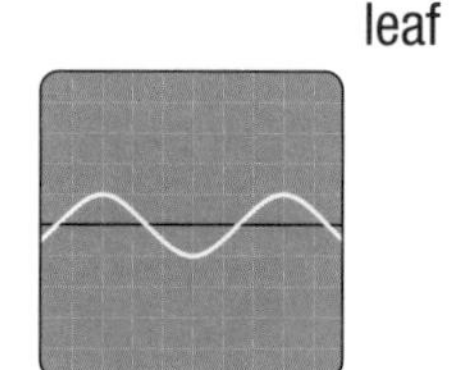

quiet sound low frequency

Top Tip!

If a dropped object smashes on the ground, its energy has been transformed from gravitational potential to kinetic to sound!

level 5

Kinetic energy

- **Kinetic** energy is the energy that all objects have when they are moving. It is sometimes called movement energy.

Did You Know?

Energy transfers are not very efficient! A car engine only transfers about one third of the stored energy in petrol into movement – the rest is lost overcoming friction in the engine and changes to heat, making the car bonnet feel warm at the end of the journey.

level 5

Electrical energy

- Electrical energy is the transfer of energy in an electric circuit.
- This electrical energy can then be used to create energy that is useful to us, for example, heat and movement energy in a hairdrier.

level 5

Types of potential energy

- The forms of energy on page 26 are all visible 'energies at work', but before an object has any of these energies it needs to have some stored or **potential energy**.
- Potential energy is the store of energy that an object has because of where or what it is. There are different kinds of potential energy: **gravitational** **elastic** **chemical**

level 5

Gravitational potential energy

- When an object is raised above the ground it gains a supply of **gravitational potential energy**. The vase on the shelf has gravitational potential energy.
- When it falls, that energy becomes kinetic energy as the vase moves towards the ground.

level 6

Elastic potential energy

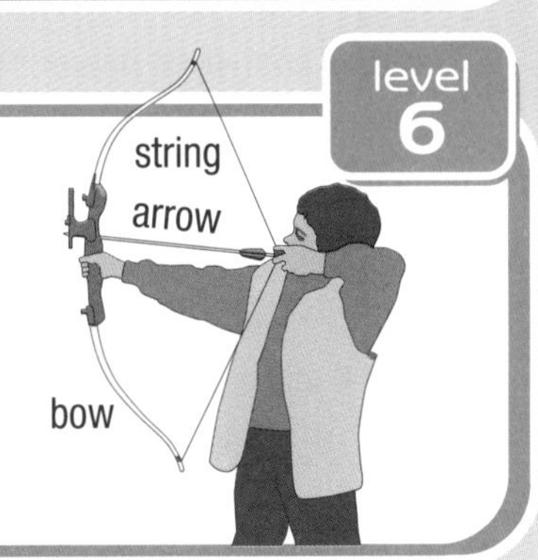

- When an elastic band or a bow string is stretched it gains a store of **elastic potential energy**.
- When the bow is released the energy becomes the kinetic energy of the moving arrow.

level 6

Chemical potential energy

- **Chemical potential energy** is the store of energy inside the petrol you put in the car or the food you eat. A candle stores chemical potential energy which is then transformed into heat energy and light energy and transferred to the surroundings when the candle burns.

Spot Check

1. What name is given to all forms of stored energy?
2. What form of energy is always part of every energy transfer?
3. What is the biggest difference between the transfer of light energy and of sound energy?

Energy transfer

Transferring energy

level 5

Energy can never be destroyed – it is just **transferred** from place to place. Sometimes it changes its form – and sometimes, when it is transferred, it is not in the form we want.

Everyday energy transfers

levels 5-6

- When we switch on an electric light, the energy has been transferred from a power station to our homes by the National Grid. The energy is then transferred to the light bulb and eventually to the surroundings.
- It is sometimes easier to think of the energy as changing its form. For example, in a light bulb it changes from electrical energy to the heat and light given out to the surroundings.

Sankey diagrams

levels 5-6

- We can show these energy transfers on a diagram as a branching arrow. These are called Sankey diagrams. **Joules** are the units of energy.

For a coal-fired power station:

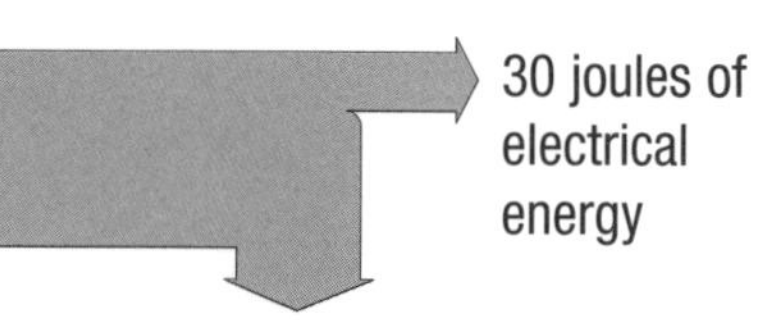

For an electric light bulb:

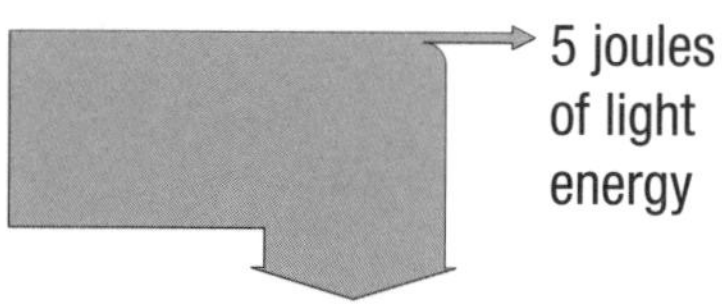

- Whatever means is used to transfer energy, there is always some energy lost to the surroundings as **thermal energy** (heat) – even if this is not the energy transfer that we want.

Top Tip!

Remember that energy is never used up. For example, when a battery is flat, it means that its store of energy has been transferred elsewhere, and then lost to the surroundings.

Thermal energy transfer

levels 5-6

- There are a lot of situations where we want to transfer thermal energy. Thermal energy can be transferred in three different ways: **conduction** **convection** **radiation**

Did You Know?

The 'radiators' in our homes and schools actually transfer heat by setting up convection currents in the air around them – rather than by radiation.

levels 5-6

Conduction

- Conduction is the transfer of thermal energy through solids.
- Metals are good conductors of heat, non-metals are not. This is why saucepans are made of metal – to transfer the heat to the food. You should stir hot things with a wooden, not a metal, spoon because wood is not a good conductor of heat so you will not burn your fingers.
- Thermal energy makes the molecules move faster. As the faster molecules collide with slower moving molecules, the energy is passed through the material.

Faster moving molecules collide with slower moving molecules.

levels 5-6

Convection

- Convection is the movement of heat through liquids and gases.
- The particles in the hot liquid (or gas) move faster and further away from each other. This makes the hot liquid less dense so it rises and the cold, more dense, liquid falls down to take its place. This process produces convection currents.

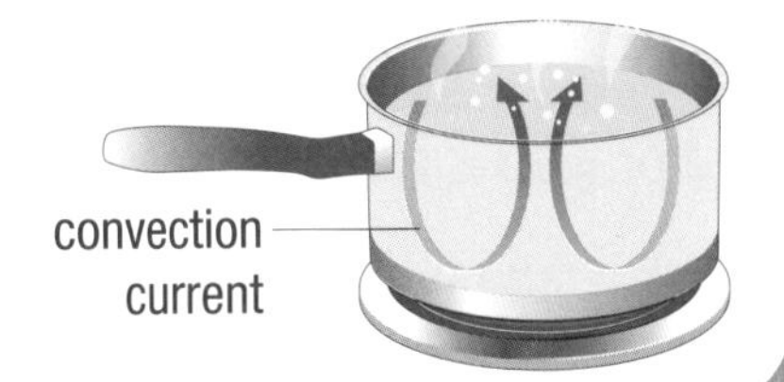

levels 5-6

Radiation

- Radiation is the heat given off by any object that is hotter than its surroundings. Black objects radiate heat more than white or silver ones. It is the way heat can move through a vacuum and is how the heat from the Sun reaches the Earth.

levels 5-6

Mixed heat transfers

- Sometimes heat transfer can be by more than one method. In the diagram on the right, heat is conducted through the cup wall and is radiated to the surroundings. Warm air above the cup rises by convection after being heated by the drink.
- You need to be able to explain what happens to the particles in a solid, a liquid or a gas during heat transfer by convection or conduction.

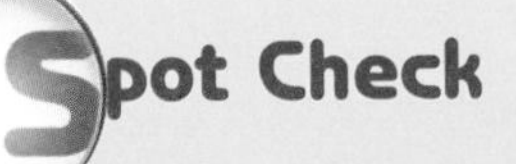

Spot Check

1. Of every 100 J of energy supplied to a hairdrier, 30 J of energy goes to the kinetic energy and sound energy of the fan. What are the other 70 J transferred as?
2. What happens to the particles in a piece of metal when it is heated at one end?
3. By what method is thermal energy transferred through liquids and gases?

Energy from food

level 4

Capturing the Sun's energy

All the Earth's energy from food comes from the Sun. The Sun's energy is trapped by plants in photosynthesis (see page 16).

Top Tip!

A green plant always starts every food chain because only green plants can trap the energy from sunlight. Animals need to eat plants, or other animals that have eaten plants, so they can get the energy they need.

level 5

Energy transfers from food

- The food we eat is a store of chemical **energy**. The energy is released by the chemical reaction of **respiration** which happens in every cell in our bodies.
- We can show respiration in a word equation like this:

Glucose + **oxygen** → **carbon dioxide** + **water** + **energy**

- This energy is then used by our bodies as heat (thermal energy) and for movement (kinetic energy) and for all the processes that living things carry out (see page 15).

green plants trap energy from sunlight

animals get energy from green plants or from other animals

- The amount of food we need to take in depends on the amount of energy we are going to use. Surplus food is stored by our bodies – and this is when we put on weight!

level 5

Food labels

- Food labels always say how much energy is in a particular foodstuff. The one shown here is from a breakfast cereal. The energy content is given in **kilojoules** (kJ) and **kilocalories** (kcal).
- Energy values are given per 100 g – so that you can compare different foods – and for a normal serving, in this case 30 g of cereal with 125 ml of semi-skimmed milk.

Nutrition

Typical values	Per 30g with 125ml semi-skimmed milk	Per 100g
Energy	738kJ **174kcal**	1624kJ **383kcal**
Protein	**5.8g**	**5.5g**
Carbohydrate	**31.7g**	**84.8g**
of which sugars	17.8g	38.3g
Fat	**2.7g**	**2.4g**
of which saturates	1.7g	1.4g
Fibre	**0.6g**	**1.9g**
Sodium	**0.6g**	**1.4g**

level 5

Different energy needs

- The recommended daily kilocalorie intake for women is 2000 kcal. For men it is 2500 kcal.
- Growing children, pregnant women, athletes and those who do hard, physical work need to take in more energy than adults who sit in an office all day.
- By carrying out an experiment to burn different foods and using the energy to heat water, we can see that different foods have different amounts of energy. A high energy food (high calorie value) will raise the temperature of water more than a low energy food.

level 5

Food groups

- Different food groups have different uses in our bodies:

Food group	Examples of foods that are good sources	What the body uses this for
Protein	Meat, fish, eggs, cheese, milk and other dairy products	Growth and repair
Carbohydrate	Bread, cake, rice, pasta, potatoes	Energy
Fat	Butter, cheese	Energy, especially long-term energy stores.
Fibre	Vegetables, wholemeal bread, cereal	Helps the movement of digested food and encourages good bacteria to grow. Not digested – passes out of the body as part of solid waste.
Vitamins and minerals	Fruit and vegetables	Lack of these causes disease.

Did You Know?

Plants trap energy from sunlight and that is how we get our energy – but they only trap about 1% of the huge amount of energy that comes from the Sun.

levels 5-6

Matching diet and lifestyle

- The amount of each type of food we eat depends on the kind of life we lead. A healthy, balanced diet contains food from all the food groups. Food is our source of energy – energy that came originally from the Sun.
- To match examples of food intake to the right people, think about their needs. A pregnant woman or a young child will need extra protein for growth. A marathon runner will need extra carbohydrate to provide energy for running.

Spot Check

1. Why is there always a green plant at the start of every food chain?
2. How would you carry out an experiment to compare the energy values in different foods?
3. Which food group that is essential to a healthy balanced diet is not digested by the body?

ENERGY Sound and light

Sound

level 6

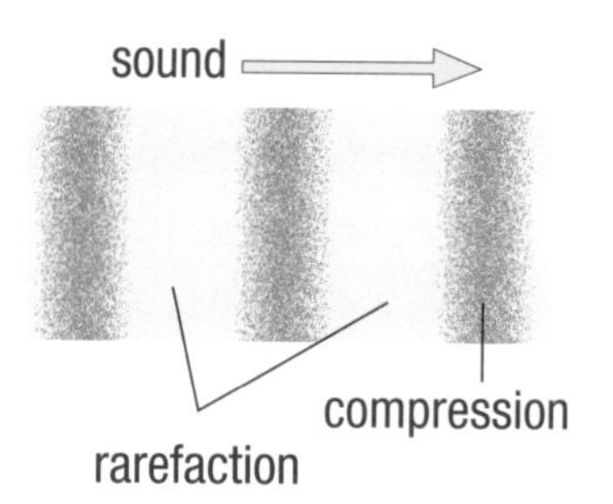

- Sound is a wave that transfers energy from one place to another. Sound travels by making particles vibrate in a system of **compressions** and **rarefactions**. Where there are no particles, sound cannot travel.

Hearing sound

level 4

- We hear sounds because our eardrum vibrates and passes the **vibration** through to the inner ear where nerves pass the impulses to the brain. As we get older we lose the ability to hear very high-pitched and very low-pitched sounds.

Top Tip!

Remember that the amplitude of a sound wave is related to the loudness of a sound and its frequency is related to pitch.

Sound waves

levels 5-6

- Sound can be described as a wave. What the wave looks like shows us what the sound will be like. The **pitch** of a sound depends on the **frequency** of the waves – that is how close together the waves are or how many waves pass a given point in one second. The loudness of a sound depends on the **amplitude** or height of the wave.

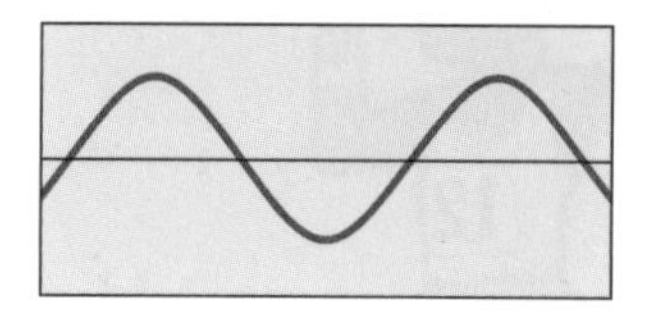

- The sound on the left would have a lower pitch than the sound on the right.

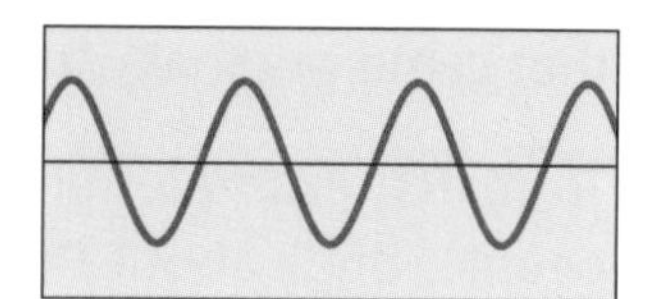

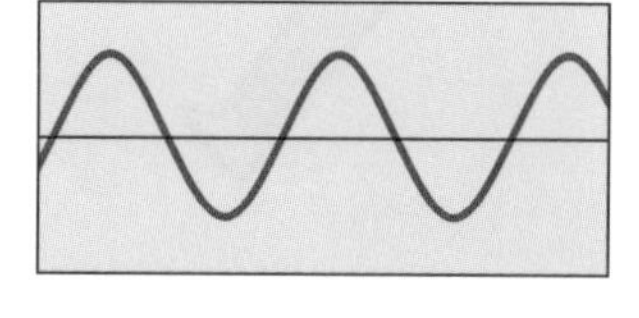

- The sound on the left would be quieter than the sound on the right.

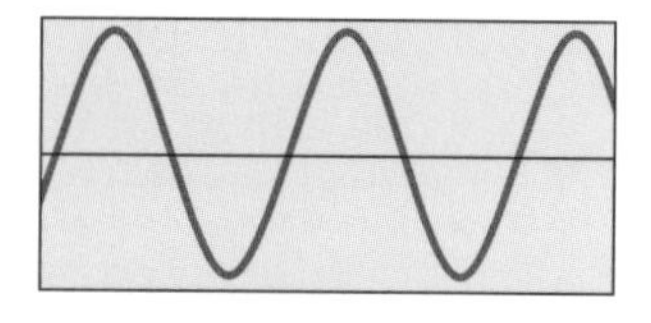

Spot Check

1. If two sound waves have the same amplitude but different frequencies, in what way will the sounds be the same and in what way will they be different?
2. Why can't we see objects in a completely dark room?
3. Why do we hear thunder after we have seen the lightning?

level 4

Light

- Light also travels as a wave – but unlike sound it does not depend on vibrations and does not need a material to travel through.

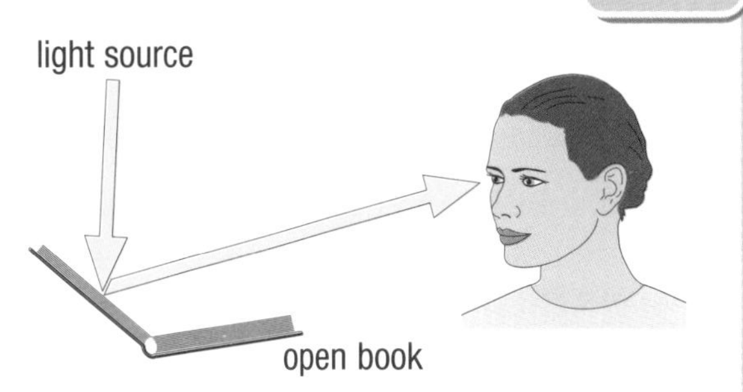

- A few things give off their own light, for example, stars (but not planets), light bulbs, candles and fires. These are called **luminous** objects.
- We see everything else because the light is **reflected** into our eyes.

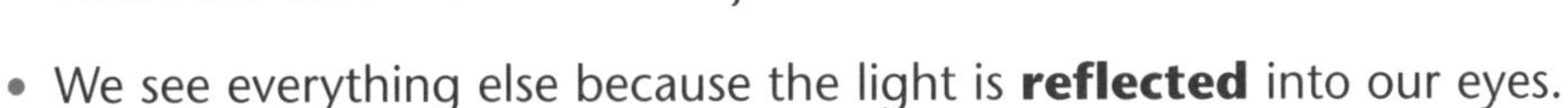

Did You Know?

'Laser' stands for Light Amplification by Stimulated Emission of Radiation. Lasers are single colour, concentrated beams of light. High power lasers can be used for cutting through metal.

Seeing straight

- Light travels in straight lines. Light can only go through materials that are **transparent**. We cannot see through brick walls because they are **opaque**. Some things, like frosted bathroom windows, are **translucent** which means they will let some light through.
- Light cannot go through opaque objects and, because it travels in straight lines, it cannot go round them. This explains why opaque objects make shadows.
- We can only see things when reflected light reaches our eyes – and light travels in straight lines. The cyclist and the car driver will only be able to see each other when there is a straight line between them and nothing opaque in between – that is when the car gets to point B.

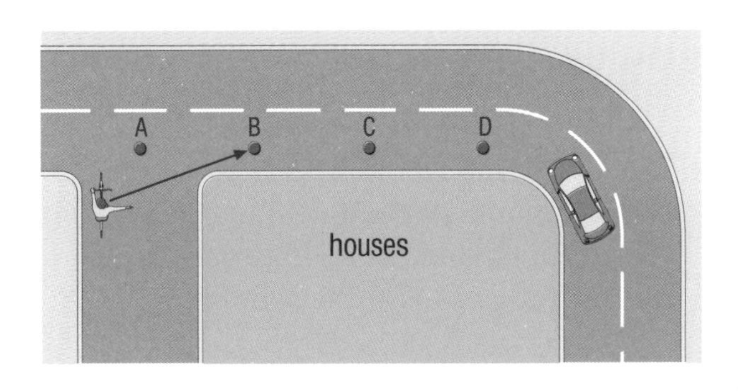

level 4

Comparing light and sound

- Sound will only travel where there are particles to vibrate. Light does not need particles to travel.
- Light will only travel in straight lines and can travel through air, a vacuum or transparent materials. Sound cannot travel through a vacuum as there are no particles.
- Light travels about a million times faster than sound. We always see lightning before we hear thunder. The gap gives us an idea of how far away the storm is – about 1 km away for every 3 seconds you count.

Reflection, refraction and seeing colour

Reflection

level 4

When light hits an opaque surface, some of it is reflected back into our eyes and we see the object. When light hits a shiny surface all of it is reflected and we see an image, such as when we look at ourselves in a mirror.

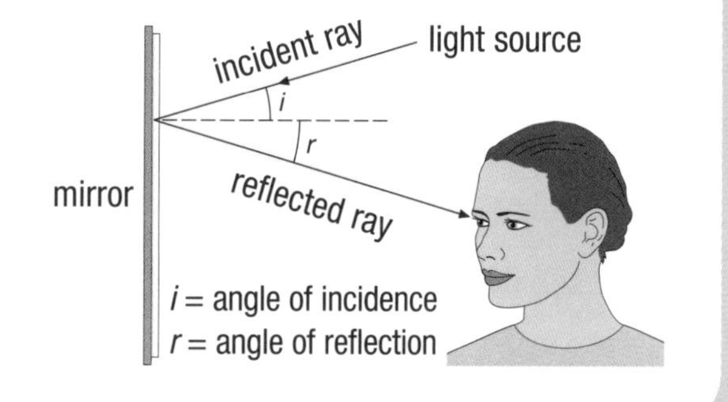

Rays

level 4

- The light rays hitting the mirror are called **incident** rays and those coming out again are called **reflected** rays.
- The angle at which the light hits the mirror is called the angle of incidence.
- The angle at which the light is reflected is called the angle of reflection.

  The angle of incidence = the angle of reflection

  This means that the light leaves the mirror at the same angle as it hits the mirror.

- We can use this fact to help us to see over high walls and around corners using a **periscope**. The light coming into the periscope hits the first mirror at 45 degrees and so is reflected out at 45 degrees which means the light has turned through 90 degrees. The same then happens at the second mirror.

Refraction

level 5

- Light travels in straight lines but it does change direction if it hits the boundary between two different materials at an angle. This is called **refraction**.
- Refraction explains:
  - why water always looks shallower than it is
  - why it is hard to tell where an underwater object really is
  - why a straw in water looks bent.

incident ray
i = angle of incidence
r = angle of refraction
refracted ray
glass block
emergent ray

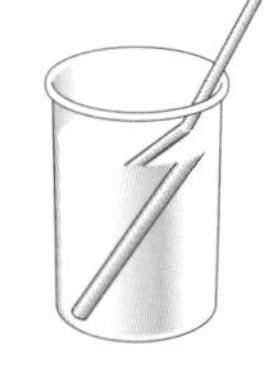

Did You Know?

Sir Isaac Newton was the first person to look at white light through a prism and see it split into the colours of the spectrum. When we see a rainbow, white light from the Sun is being split into colours because raindrops in the air act as prisms.

level 6

Seeing colour

- White light is a mixture of the seven colours of the spectrum. White light can be split into a **spectrum** using a **prism**.

spectrum

Top Tip!

The colours of the spectrum, Red, Orange, Yellow, Green, Blue, Indigo and Violet can be remembered like this: Richard Of York Gave Battle In Vain.

levels 6-7

Reflecting colour

- We see things as being coloured because, when the white light falls on them, only part of the spectrum is reflected into our eyes.
- Red objects reflect red light; blue objects blue light, etc. White objects reflect all the colours of light. Black objects look black because they absorb all the light and do not reflect any of it.
- This means that a blue object in red light will look black. This is because blue objects can only reflect blue light, they absorb all the other colours. If there is no blue light then nothing is reflected and so the object looks black as it absorbs all the light – just like a black object.

level 7

Mixing coloured light

Coloured beams of light can be mixed to make light of different colours.

- The primary colours of LIGHT are red, green and blue.
- An object that looks yellow is reflecting red and green light. In red light a yellow object would look red and in green light it would look green but in blue light it would look black as it would not be able to reflect any of the light falling on it.

Red
Yellow
Green
White
Magenta
Cyan
Blue

level 7

Filters

- A coloured filter only allows its own colour of light to pass through and it absorbs the rest of the spectrum. A red traffic light is a bulb giving off white light behind a red filter that only lets red light through.

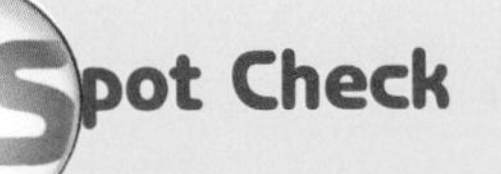

1. When white light is split into a spectrum by a prism, which colour of light is bent the most?
2. What colour would an object that looks yellow in white light look in red light?
3. Why does an object that looks white in white light look red in red light?

Balanced and unbalanced forces

level 4

Forces acting on an object

- You need to think about all the forces acting on an object and how they combine to make the object move or stay still.
- The unit for all forces is the **newton**. Weight (but not mass) is a force and so is measured in newtons (N).
  - A mass of 100 g has a force of about 1 N due to gravity.
  - A mass of 1 kg has a force of about 10 N due to gravity.

Top Tip!

You need to understand the information on these two pages to be able to answer higher level questions on forces.

level 4

Balanced forces

- Look at the picture of a man standing on a bench.
- The weight of the man pushes down on the bench with a force of 750 N. The bench pushes back up with a force of 750 N. The forces are **balanced**.
- Stationary objects have balanced forces acting on them. They do not move because the forces are balanced NOT because there are no forces at all – forces are acting on everything all the time.
- If an object is moving and the forces on it are balanced, then it will continue to move at the same speed in the same direction until something happens to make the forces **unbalanced**.
- This boat is floating because the weight of the boat acting downwards is exactly balanced by the upthrust from the water acting upwards. The forces are balanced.

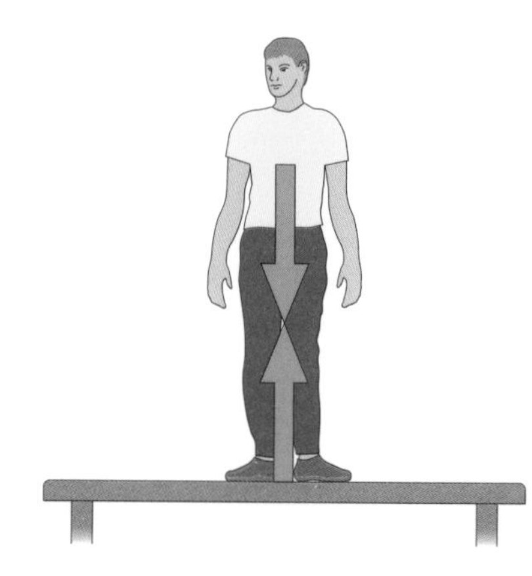

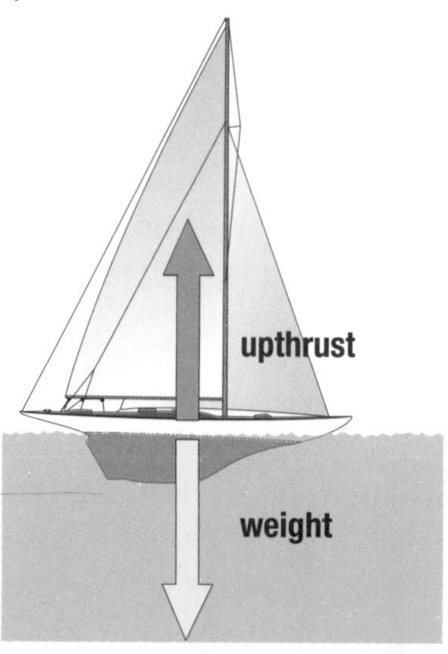

Did You Know?

The line painted on the side of cargo ships is called the Plimsoll line – after Samuel Plimsoll. When fully loaded, a ship can safely go down in the water to this level. Upthrust from the sea will balance the load and the ship will float. The temperature of the water and its salinity determine the level of the line.

levels 3-4

Unbalanced forces

- Unbalanced forces make things change:

speed **direction** **shape**

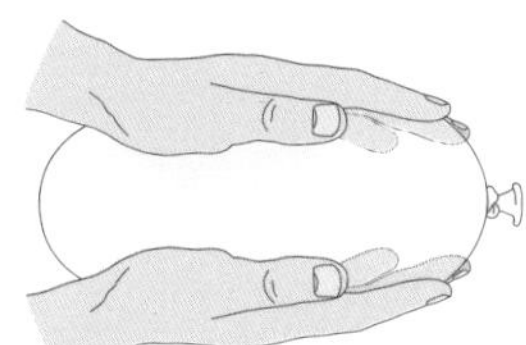

Pushing a bowling ball
The **push** force makes the ball move as it is greater than the other forces, such as friction, which act on the ball.

Squashing a balloon
The balloon is squashed and changes shape because the push of the hands is greater than the force of the air acting on the inside of the balloon.

Twisting a bottle top
The **twist** force on the bottle top is greater than the friction, so the top starts to move.

Pulling a water skier
The **pull** force of the boat on the skier is greater than the drag of the water, so the skier moves forward.

level 5

Speeding up and slowing down

- For a car to increase its speed, the force from the engine (using the accelerator) must be greater than the **air resistance** and the **friction**, which act to slow it down.
- To make a car reduce its speed, the forward force from the engine must be reduced by not pressing the accelerator pedal and the friction increased by putting on the brake. There is more about air resistance on page 41 and about friction on page 38.

level 5

Pressure

- The pressure arising from a force acting on any object will be greater if the force acts over a smaller area. That is why knife blades and the points of needles are very narrow – to make the pressure as great as possible.
- Skis and tractor tyres have the largest possible area to spread the force and make the pressure as low as possible to stop them sinking into the ground.

Spot Check

1. What **three** things can an unbalanced force change in a moving object?
2. If the forces on a moving object are balanced, what will happen to the moving object?
3. How can you increase the pressure on an object without increasing the size of the force?

Friction

level 5

Friction between moving objects

Wherever there is movement there is **friction** – and friction always acts against the movement causing it.

- If you rub your hands together very quickly they will get hot. If you slide on a carpet you may get a carpet burn. These are both everyday examples of friction.
- Whenever two solid touching surfaces move, there is friction. The rougher the surfaces, and the faster they move, the more friction there will be.

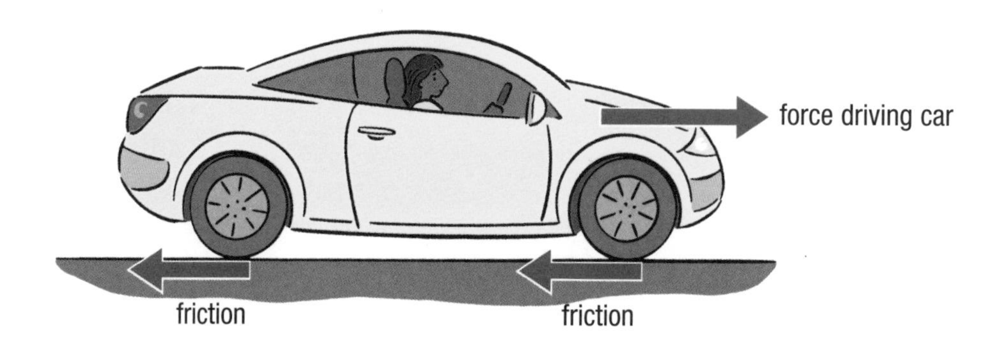

level 5

Balancing friction

- In this diagram, it is the friction between the chain and the table that is stopping the chain from falling off the table. The two forces of weight and friction are balanced.

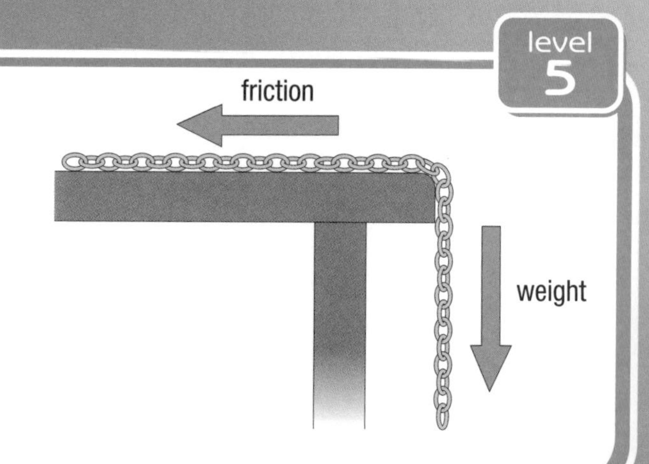

Top Tip!

Don't forget friction when thinking about whether forces are balanced or unbalanced.

level 5

Overcoming friction

- The force you need to apply to move this box will have to be greater than the friction between the box and the floor. If the size of the push is less than or equal to the friction, then the box will stay still.

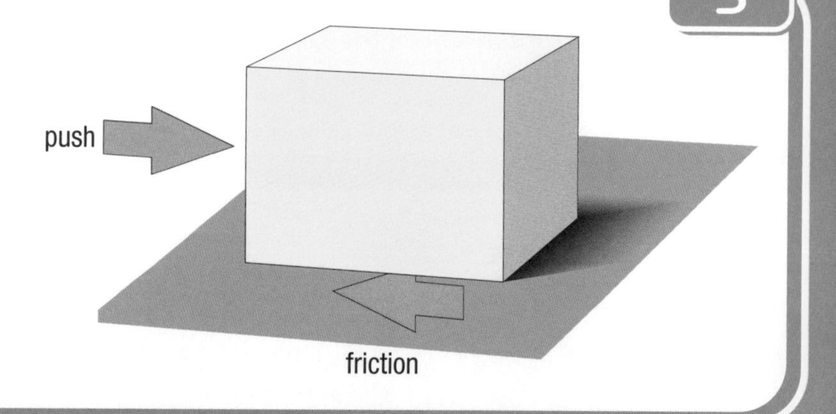

level 6

Effects of reduced friction

Friction can be a great nuisance – it causes wear on all sorts of moving parts in many types of machinery. It is also, however, something we could not manage without.

- When you slip on an icy pavement on a winter morning, it is because there is not enough friction between your shoes and the pavement.
- Cars take longer to stop on wet or icy roads because there is less friction between the tyres and the road.
- Worn tyres are dangerous because they reduce the friction between the car and the road.

Did You Know?

The first bicycles appeared in Paris in 1791. By the 1880s they had become very fashionable, especially the penny farthing with its large front wheel and a small back wheel.

level 6

Using friction

- Think about all the points where friction is involved when riding a bike:

The brakes rely on friction to slow the bike down and make it stop.

For the wheels to turn easily the axles need to be oiled to reduce friction.

The friction between the tyres and the road give grip which pushes the bike forward.

Friction keeps your feet on the pedals.

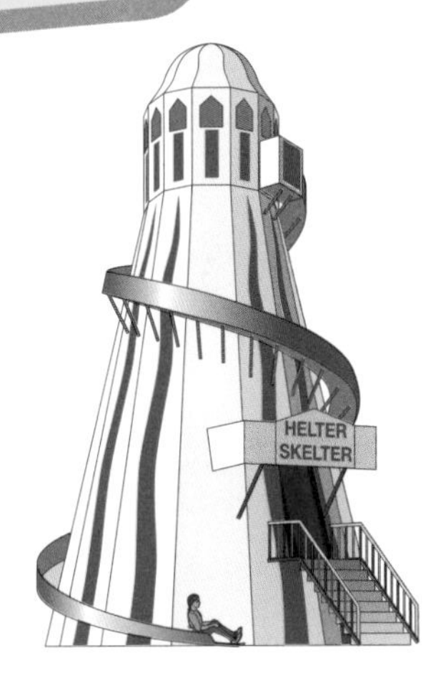

- On a slide, friction will slow you down, which may or may not be a good thing. A slippery mat will help you go faster because it will reduce the friction between your body and the slide.

Spot Check

1. Read these statements carefully about the friction acting on a moving car. Which ones are examples of **useful** friction?
   - **a** The friction between the tyres and the road.
   - **b** The friction between the moving parts of the engine.
   - **c** The friction between the brake blocks and the wheels.
   - **d** The friction between the wipers and the windscreen.
   - **e** The friction between the driver's hand and the steering wheel.
2. A person is trying to move a heavy block that is resting on the ground by pushing it. What will be the size of the push force compared to the friction force at the point when the block starts to move?

Streamlining and air resistance

levels 5-6

Friction in liquids and gases

Friction does not only happen between solid surfaces. Liquids and gases exert a kind of friction on objects moving through them as well.

levels 5-6

Air resistance

- Liquids and gases both exert friction on objects moving through them. The **drag** that a car, aeroplane or a parachutist feels when moving quickly through the air is called **air resistance**.
- Air resistance increases the faster you move, until the air resistance is so great that you cannot go any faster. This is why everything has a top speed. Cars, aeroplanes and even cyclists' helmets are designed to reduce air resistance and make the top speed as high as possible.

levels 5-6

Streamlining

- If cars have a **streamlined** shape it not only makes the top speed greater, but it also reduces the size of the forward force needed from the engine to move the car at any speed. The smaller this force needs to be, the less fuel will be needed to drive the car at any speed.
- In nature, fish and sea creatures have streamlined shapes to enable them to move through water with the least possible water resistance.

Air resistance is a special example of friction. The faster you move, the more air resistance you will experience. When you run or ride a bike very fast the wind feels strong compared with when you just walk slowly. That is air resistance.

Did You Know?

Without a parachute to slow them down, sky divers accelerate at a rate of 9.8 m/s/s. This means that in each second their speed is almost 10 m/s faster than it was in the last second.

levels 5-6

Air resistance and unbalanced forces

- When a parachutist jumps out of a plane, the force of gravity (see page 48) makes him accelerate towards the ground, and air resistance acts upwards to slow him down. The forces on the parachutist are unbalanced as the air resistance is less than the force of gravity and this causes him to fall faster.
- When arrows are used in diagrams to represent forces, the direction of the arrow shows the direction of the force and the size of the arrow shows the size of the force.

level 6

Terminal velocity

- Eventually the parachutist will be moving so fast that the air resistance slowing him down will be as great as the force of his weight speeding him up. As the forces are then balanced, he will stop accelerating and continue to move at a **constant speed**. This speed is known as **terminal velocity**.
- The upward force from the air resistance is now the same as the downward force caused by the parachutist's weight. The arrows showing the two forces are now the same size.

level 6

Changing forces

- When the parachutist opens his parachute, the air resistance will become much greater. This will slow him down. As his speed reduces, the air resistance will then become less. Eventually the forces will balance again and he will stop slowing down and continue at a new – but slower – terminal velocity.

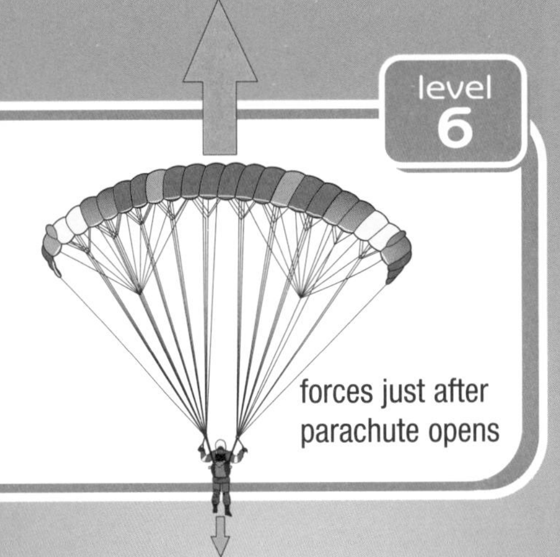

Spot Check

1. What is the name of the force that causes everything to have a maximum possible speed?
2. What name is given to that maximum speed?
3. Which of the forces acting on a parachutist does not change at any point as they fall through the air?

Speed and motion

levels 4-5

Speed

- If a car travels a **distance** of 50 kilometres in one hour, we say it has travelled at a **speed** of 50 kilometres per hour (km/h). If a person walks 10 metres in 10 seconds we say they have walked at a speed of 1 metre per second (m/s).
- If we know the distance travelled and the **time** taken, we can calculate speed using the equation:

  Speed = distance ÷ time

  If a car travels 200 km in 4 h then:

  Speed = distance ÷ time
  = 200 ÷ 4
  = 50 km/h

- The units used for speed not only need to be correct, but also suitable for the purpose. Speeds are generally measured in metres per second (m/s) but for a car or an aeroplane, kilometres per hour (km/h) or miles per hour (mph) is better. For something moving very slowly then it might be millimetres per second (mm/s) or even millimetres per hour (mm/h).

levels 4-5

Distance and time

- As well as calculating speed, you can use the same equation to work out distance if you know the time taken for the journey and the speed.
- You can work out how much time a journey would take if you know both distance and speed.

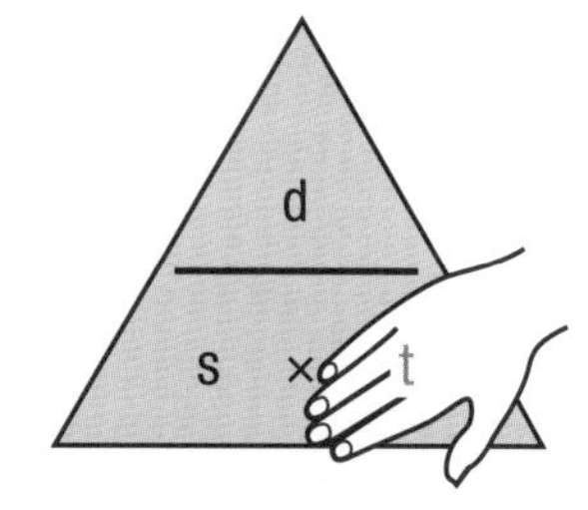

- This triangle can help you to re-arrange the equation. Just cover up the quantity you want and you will be able to see how to work it out.

Top Tip!

Always include the units with your answer when working out speed, distance or time.

levels 4-5

Average speed

- It would be most unusual for any journey to be travelled at exactly the same speed all the way. A bus stops to pick up passengers, a car slows down at traffic lights and roundabouts and even a person walking has to stop before they cross the road. It would be very complicated to work out the individual speeds for each part of any journey, which is why we use the total distance and the total time to give us the **average speed**:

$$\textbf{Average speed} = \frac{\textbf{total distance travelled}}{\textbf{total time taken}}$$

levels 4-5

Distance-time graphs

- We can show what an actual journey looks like on a distance-time graph.
- Between A and B the bus is moving at a steady speed. The line is going up and is straight.
- Between B and C it is stationary. The line doesn't go up.
- It then moves at a slow, steady speed between C and D. The line is not very steep.
- It then moves at its fastest speed for the journey – it covers more distance in the time – from D to E. Here the line is steepest.
- Between E and F it is stationary again.

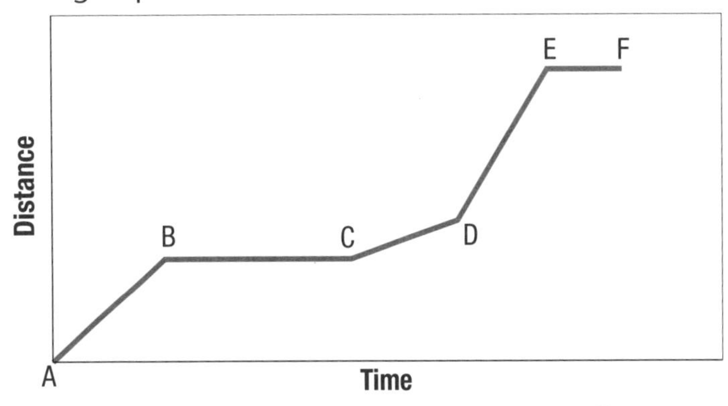

Did You Know?
The current land speed record is 763 mph.

level 7

Speed-time graphs

- A graph of speed and time will show us how the speed of the object changed throughout a journey. In these graphs …

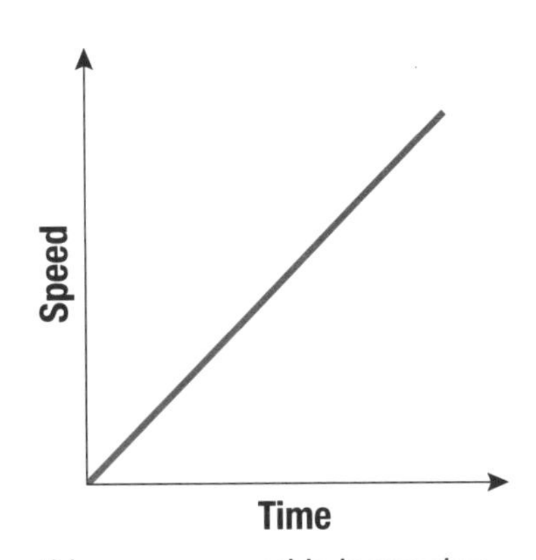

this means speed is increasing

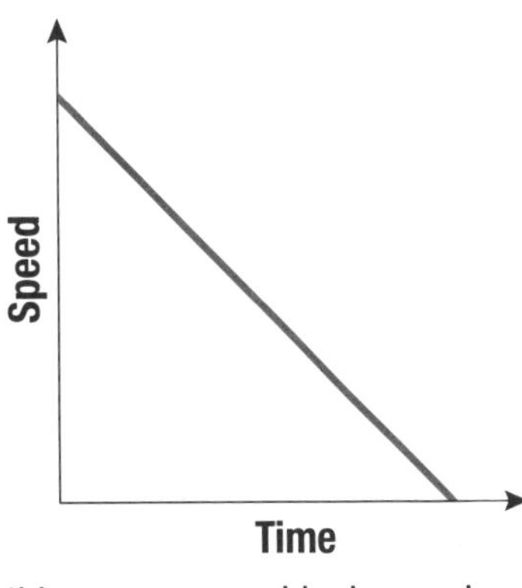

this means speed is decreasing

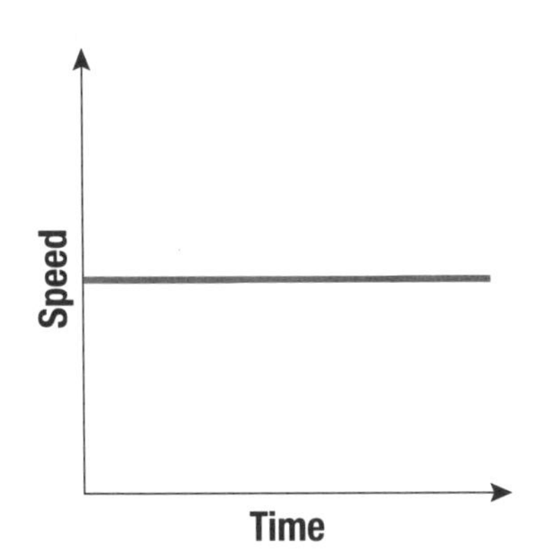

and this is how a graph, or part of a graph looks if an object is moving at a steady speed.

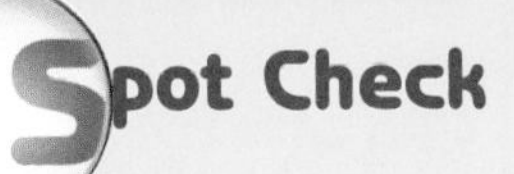

Spot Check

1. If a train journey of 150 miles takes 3 hours, what was the average speed of the train?
2. Does this mean that the train travelled at a steady speed for the whole journey?
3. What would be suitable units to describe the speed of a sprinter running a 110 m race?

Speed and stopping distance

level 7

How does speed affect stopping distance?

The faster a car is travelling, the longer it will take to stop.

- First the driver has to decide to stop.
- While this is happening the car continues to travel along the road – the distance travelled at this point is called the **thinking distance**.
- The driver then puts their foot on the brake.
- Once the brake is pressed it takes a little more time for the car to actually stop – the distance the car travels in this time is called the **braking distance**.
- The thinking distance and the braking distance added together give the total **stopping distance** for a car travelling at that speed.

level 7

Working out stopping distances

The faster the car is travelling the greater the stopping distance.

- At 30 mph (miles per hour), thinking distance is 9 metres and braking distance is 14 metres giving a total stopping distance of 23 metres.

30 mph

9 metres thinking distance | 14 metres braking distance | = 23 metres total stopping distance

- At 70 mph thinking distance is now 21 metres and braking distance is 75 metres which makes a total stopping distance of 96 metres.

70 mph

21 metres thinking distance | 75 metres braking distance | = 96 metres total stopping distance

Top Tip!

Speed, road surface, the state of the tyres and driver's reaction time all contribute to the time between seeing a hazard and stopping the car.

level 7

Friction and stopping distance

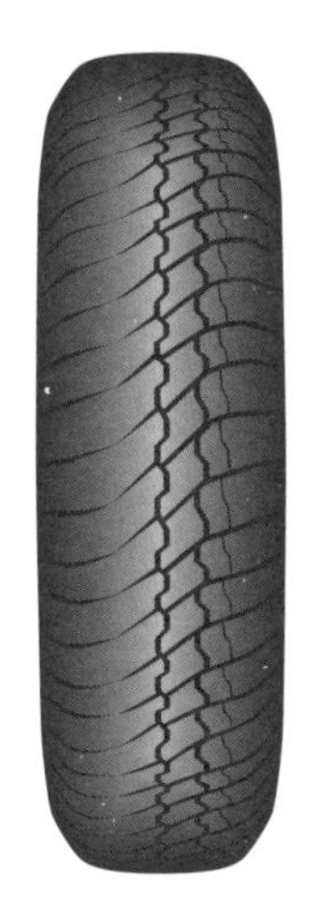
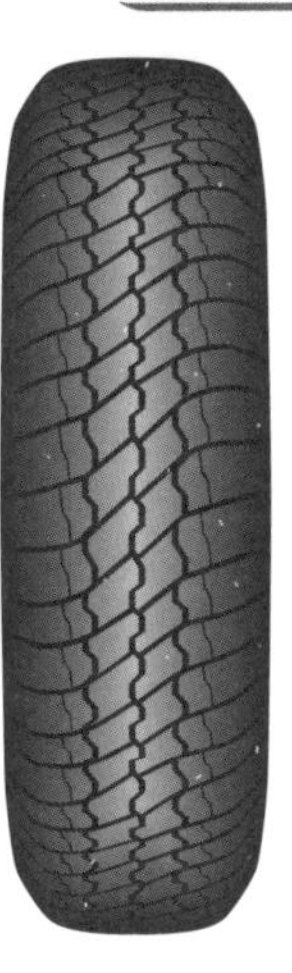

- For a car to stop quickly, the **friction** between the tyres and the road and between the brakes and the tyres must be as great as possible.
- The distance travelled by a car with the old tyre (on the left) after the brakes have been applied would be greater than the distance travelled by a car with the new tyre (on the right), because the new tyre would have a better grip on the road. There would be more friction so the stopping distance would be less.
- The road surface will also affect stopping distance for the same reason. A very smooth road surface – especially one that is very wet or icy – will have much lower friction and so the car will travel further after the brakes have been applied.
- Friction affects braking distance and therefore stopping distance.

level 7

Drivers and thinking distance

- The first part of stopping distance – thinking distance – has been travelled even before the brakes have been applied.
- The thinking distance will be greater if the driver's **reactions** are slower. This would happen if the driver was tired, under the influence of drugs or alcohol, or distracted by something such as a mobile phone.

Did You Know?

The first journey by a petrol-driven car was in July 1886 when Karl Benz drove his three-wheel vehicle for about one mile at a speed of about 9 miles per hour.

Spot Check

1. What two distances are added together to make overall stopping distance?
2. What factors might increase thinking distance?
3. Why would overall stopping distance be greater on a wet road?

Pressure and moments

level 7

Pressure

Pressure depends on two other things: **force** and **area**.

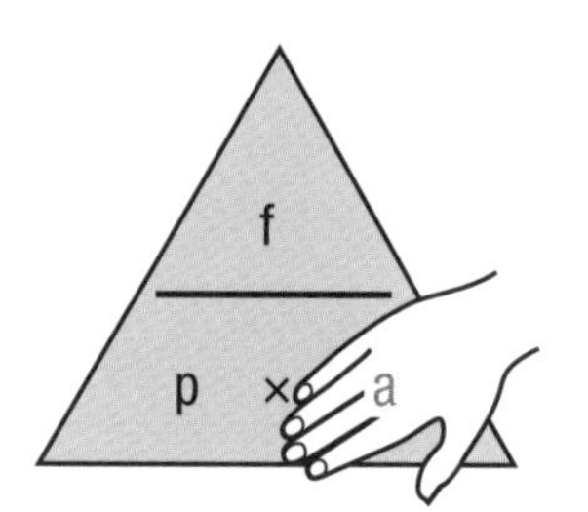

- The greater the force, the greater the pressure.
- The smaller the area, the greater the pressure (see page 37).
- To calculate pressure we use this equation:

  Pressure = force ÷ area

  You can use a triangle – as with the speed equation – to help you rearrange this equation.

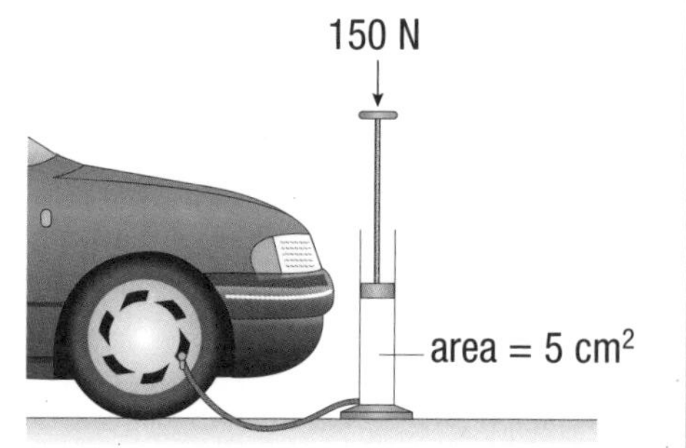

- If the piston of this pump has an area of 5 cm^2 (cm squared) and the driver of the car pushes down with a force of 150 N, we can calculate the pressure on the air in the pump.

  Pressure = force ÷ area
  = 150/5
  = 30 N/cm^2 **NOTE THE UNITS!**

- When the tyres have been pumped up and the car is driven away, each tyre exerts a pressure of 30 N/cm^2 on the road. The area of the tyre in contact with the road is 100 cm^2. We can calculate the force exerted on the road by each tyre by re-arranging the equation.

  Pressure = force ÷ area So **force = pressure × area**
  = 30 N/cm^2 × 100 cm^2
  = 3000 N

Did You Know?

Water boils at a higher temperature if the pressure is high, and at a lower temperature if pressure is low. A pressure cooker is a type of saucepan that makes use of this to cook food more quickly.

level 7

Heat and pressure

- Pressure is caused by the gas molecules moving about and colliding with the walls of the tyre or other container.
- Heating a gas gives the molecules more thermal energy, so they move more – so there are more collisions and the pressure gets greater.

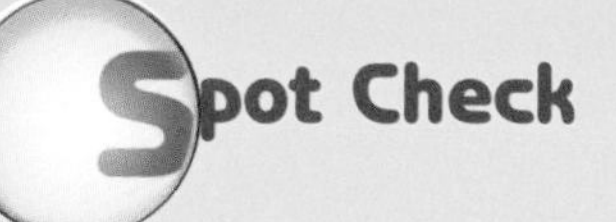

Spot Check

1. A gardener pushes down on a spade with a force of 200 N. If the surface area of the end of the spade is 10 cm^2, what pressure does he exert on his flower bed?
2. If the surface area of his boot is 350 cm^2, how much extra pressure has the spade allowed him to exert?
3. What **two** factors affect the size of a moment or turning force?

Top Tip!

As turning forces are bigger the further away they are from the pivot, spanners and pumps are made with long handles so that applying a small effort results in a big turning force being exerted.

level 7

Moments

- As well as acting in straight lines, forces can act to make objects turn. The proper name for a **turning force** is a **moment**. The size of a moment depends on two things:
  - the size of the actual force
  - the distance from the **pivot** at which that force is applied.
- If an object has equal and opposite turning forces it does not move. We say it is in **equilibrium**.

level 7

Clockwise and anticlockwise moments

Turning forces cannot be said to act just up or down or to the left or right, so we describe them as being **clockwise** or **anticlockwise**. Think about the sort of mobile toy that might hang above a baby's cot.

- The place where the string from the ceiling is attached to the bar of the toy is called the pivot.
- Shape A is acting in an anticlockwise direction and shapes B and C, which are hung on the other side of the pivot, are both acting in a clockwise direction.
- For each of the shapes in the mobile, we can work out the moment:

  moment = distance from the pivot × size of the force
  - for A: moment = 10 cm × 8 N = 80 Ncm
    A = total anticlockwise moment.
  - for B: moment = 5 cm × 10 N = 50 Ncm
  - for C: moment = 15 cm × 3 N = 45 Ncm

    B + C = total clockwise moment = 50 + 45 = 95 Ncm.

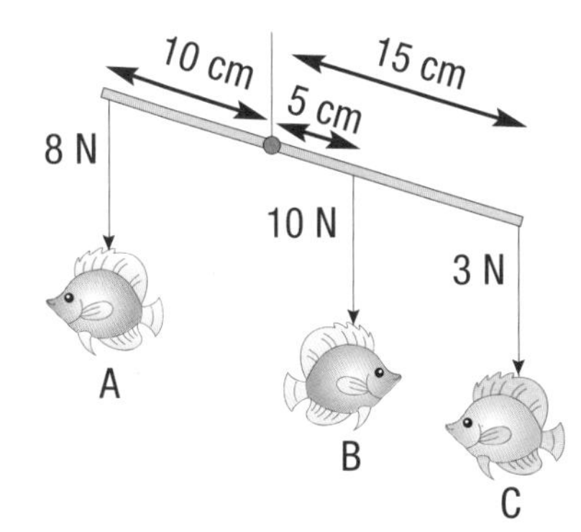

- The mobile will not balance and hang straight.

level 7

Balancing moments

- To make the mobile above balance, a turning force of 15 Ncm (95 – 80 Ncm) needs to be added to the same side as A, for example, a shape with a force of 3 N added 5 cm (3 × 5 = 15 Ncm) from the pivot.
- This principle can be applied to any situation but remember:
  - each moment needs to be worked out separately
  - if there are two or more moments acting together on one side of the pivot, add them up
  - anything balanced on a pivot will be straight if the moments on each side are the same – and will tip if the moments are different.

Gravity

level 3

Gravity, mass and weight

Like friction and air resistance, **gravity** is a force that is all around us.

- Any object that has a **mass** is subject to the force of gravity that causes a downward force called **weight**.

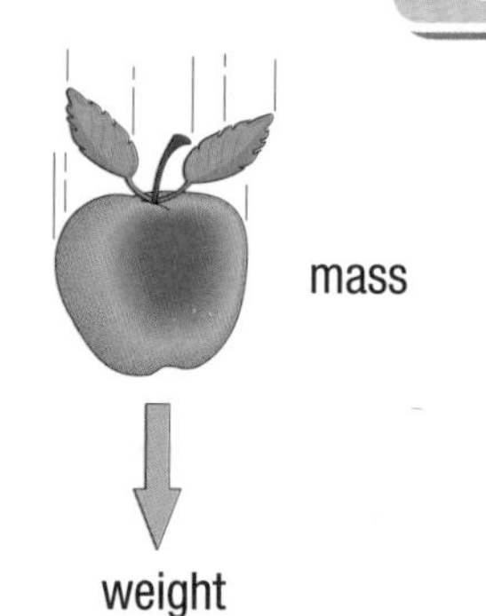

Top Tip!

Don't confuse mass and weight. Mass is the measure of how much there is of something (in kilograms). Weight is a force caused by gravity (in newtons).

level 4

Falling

- We think of things falling downwards, but in fact they are falling towards the centre of the Earth – otherwise, if people on the other side of the Earth dropped things they would float off into the sky!
- Wherever you are on the surface of the Earth, downwards always means towards the centre of the Earth.

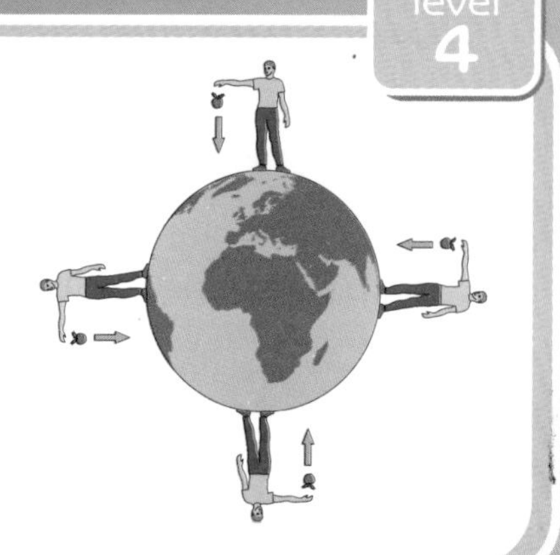

level 4

Gravitational pull

- Any two bodies exert a gravitational force on each other – but the effect of that gravitational pull depends on the mass of the body. This is why an apple falls towards the Earth – the Earth does not move towards the apple!
- On Earth a mass of 1 kg has a weight of 10 **newtons**.

Gravity and the Moon

On Earth

- The mass of the Moon is much less than the mass of the Earth. This makes the gravitational force pulling objects towards the centre of the Moon much less than the force pulling objects towards the centre of the Earth.
- This means that if we are on the Moon, although our mass is the same, our weight is less – about one-sixth of what it would be on Earth.

On Moon

level 5

Gravity on other planets

- On a planet with a bigger mass than the Earth, the gravitational pull is greater, so an object weighs more, but it still has the same mass.
- The amount of gravitational pull depends on the distance from the object as well as on the mass of the object. So if you travel in space, far from the planets, you will not be affected by the gravitational pull from any of them.

level 5

Newton and Galileo

- Forces are measured in units called newtons after Sir Isaac Newton, a scientist who lived from 1642 to 1727. He is said to have 'discovered' gravity when an apple fell on his head! Newton worked out how to calculate the gravity acting on objects – and used his ideas about gravity to predict how the planets moved around the Sun.
- Before Newton, an Italian scientist called **Galileo** Galilei did experiments on the force affecting falling objects. He thought that if he dropped objects of different masses from a great height, they would all reach the ground together. His theory was right, but the objects all hit the ground at different times because he did not take into account the air resistance acting on each object.
- Hundreds of years later Galileo's experiment was repeated on the Moon using a hammer and a feather. As there is no atmosphere on the Moon, there is no air resistance and so both objects hit the ground together.

Did You Know?

As well as his more famous scientific work, Isaac Newton invented the cat flap!

1. If a person has a mass of 70 kg, what would their weight be: **a** on Earth; **b** on the Moon?
2. You want to show someone that Galileo was right – without going to the Moon. What other factors would you have to keep the same if you dropped a heavy object and a light object from the same height at the same time to show that they hit the ground together?

FORCES The Earth and beyond

level 3

The Earth and the Moon

Newton's ideas about gravity and forces help us to understand the motion of the **Moon** and the planets.

- The Moon is much smaller than the **Earth**, the Earth is much smaller than the **Sun** – so the Moon is very, very much smaller than the Sun. When we look at the sky, the Moon and the Sun look about the same size. This is because the Sun is much further way. The Sun is about 400 times bigger than the Moon but it is about 400 times further away.

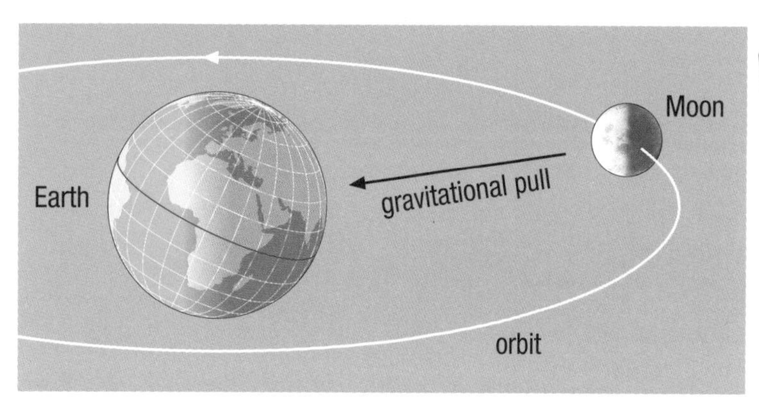

Did You Know?

Mercury, Venus, Mars, Jupiter and Saturn were discovered long ago, Uranus in the 1780s, and Pluto and Neptune even more recently. New smaller bodies are being discovered in space all the time. In 2003, astronomers discovered a tenth planet that they called Sedna.

- The Moon **orbits** the Earth about once every 28 days – the Moon is kept on its path by the action of the Earth's gravitational field. The shape of the path is called an **ellipse** – rather like the shape of an egg.
- The Moon does not give off its own light but is illuminated by the Sun and reflects the Sun's light towards the Earth. The amount of the lit side of the Moon that is visible from Earth changes; when the Moon is between the Earth and the Sun then very little of the lit side of the Moon can be seen from Earth and we call this a new moon. When the Moon is on the far side of the Earth from the Sun then the whole of the lit side can be seen and we call this a full moon. In between, we see crescent moons and half moons.

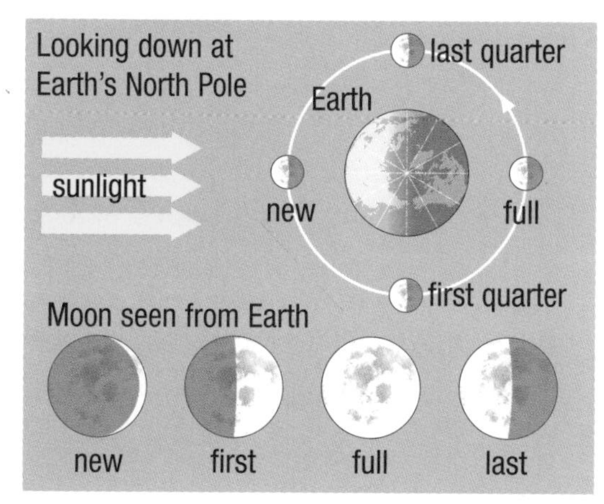

The Earth and the Sun

- The Earth orbits the Sun in the same way as the Moon orbits the Earth. Long ago people thought that the Earth was at the centre and that the Sun and the other planets travelled around the Earth. Around 1500, a Polish astronomer called Copernicus was the first to suggest that the Sun was at the centre. Galileo agreed with this theory when he used an early telescope to observe the planets.

level 4

Day and night

- The Earth spins on its **axis** once in 24 hours – this gives us day and night.

level 4

Summer and winter

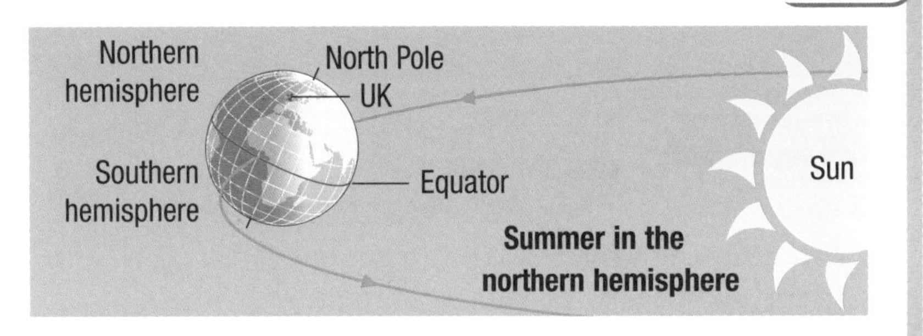

- The Earth orbits the Sun once in $365\frac{1}{4}$ days. The 365 days make one of our years and the $\frac{1}{4}$ days are added to give us an extra day in February every 4 years. In the course of a year we have summer and winter because the Earth is tilted on its axis. When the northern **hemisphere** – where Britain is – is tilted towards the Sun, we have summer. When it is tilted away from the Sun, we have winter.

Top Tip!

Remember that the Earth spinning on its axis gives us day and night, but its orbit around the Sun determines our seasons.

The solar system

- The Earth is not the only planet to orbit the Sun. There are eight other planets in our **solar system** as well. It is important to remember the order of the planets.

Planets in order of distance from the Sun	MERCURY	VENUS	EARTH	MARS	JUPITER	SATURN	URANUS	NEPTUNE	PLUTO
A way of remembering	MY	VERY	EASY	METHOD	JUST	SPEEDS	UP	NAMING	PLANETS

- The movements of the Earth and the other planets around the Sun follow the same kind of elliptical (egg shaped) paths as the Moon around the Earth. These paths are ellipses rather than circles because as a planet gets nearer to the Sun it is pulled closer by the huge gravitational field.
- The length of a year on a planet varies depending on how far away it is from the Sun as the more distant planets take much longer to travel all the way around the Sun.

1. Which **two** planets will have a shorter year than an Earth year and why?
2. Over the centuries our ideas of how many planets there are in the solar system has changed, why has this happened?
3. What does this tell us about how scientists think and work?

Magnetic forces

Magnetic fields

level 4

Like gravity, magnetism is a force that acts at a distance. Unlike gravity, magnetic forces can repel as well as attract.

- A bar **magnet** has a magnetic field all around it. Iron filings will follow the **magnetic field** lines and give a pattern around a magnet.

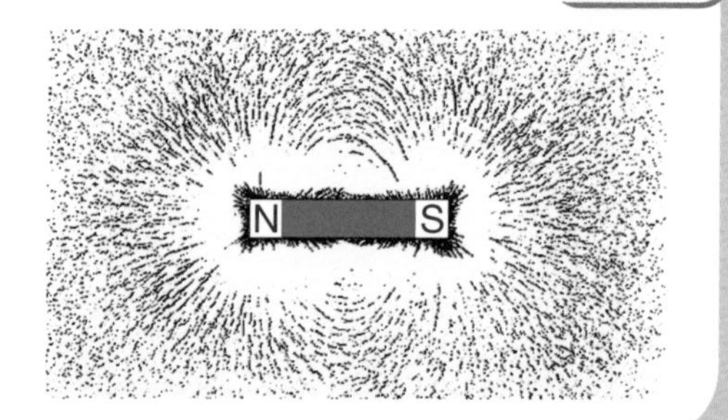

Dipoles

level 4

- Inside a piece of magnetic material such as iron, the atoms all behave like tiny magnets, called dipoles. If the iron has not been magnetised then all these dipoles are randomly arranged.
- When the iron becomes magnetised, all the dipoles line up with their North **poles** facing in one direction and their South poles facing the other way.
- The dipoles can all be lined up to create a magnet by stroking the iron with an existing magnet or by using the magnetic field produced by an electric current (see page 53).

An unmagnetised piece of iron

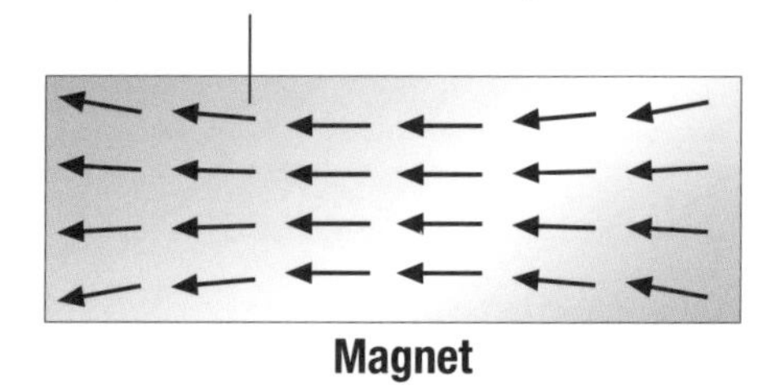

Magnet

Repulsion and attraction

levels 4-5

- If you put two magnets close to each other, and two like poles (North-North or South-South) are facing each other they will repel (this means push away). If two unlike poles (North-South) are together, they will attract.
- These two magnets will move away from each other …

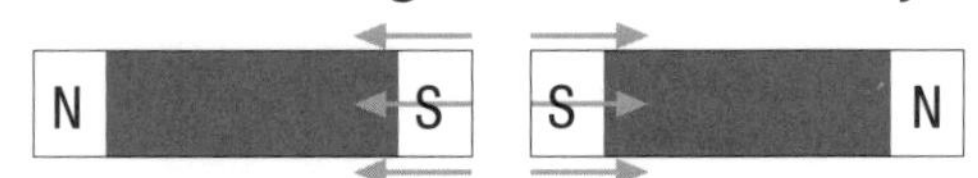

… but these two will move together.

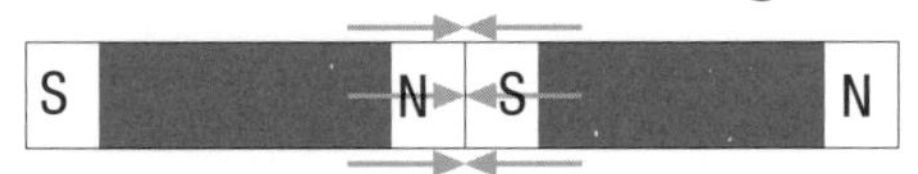

- A magnet will attract a piece of magnetic material, but only two magnets can repel each other.

Did You Know?

The strongest magnet in the world is an electromagnet in California which is 250 000 times stronger than the Earth's magnetic field.

level 5

The Earth's magnetic field

- The Earth has a magnetic field – in fact it behaves as if it has a giant bar magnet in its centre.
- The North-seeking pole of a magnet inside a plotting compass will point to magnetic North, which is not in quite the same position as the geographic North Pole.

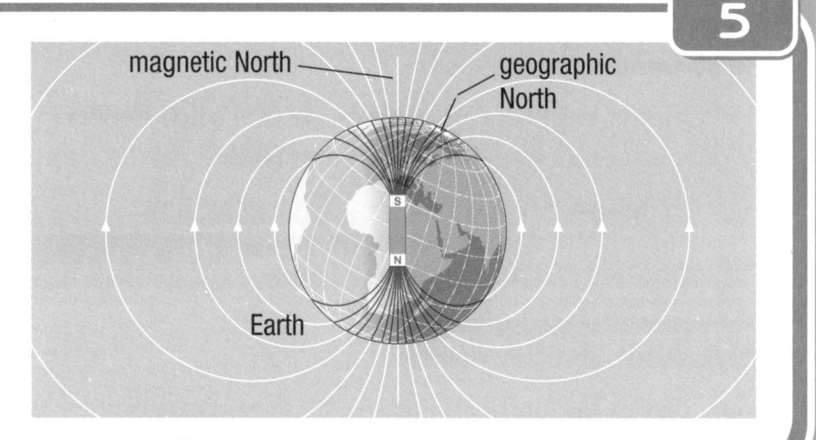

Top Tip!

The only magnetic elements are iron, nickel and cobalt. Steel is magnetic because it is made mostly from iron.

levels 5-6

Electromagnets

- Whenever a current flows in a wire it creates a magnetic field around it.
- This magnetic field can be used to magnetise an iron **core** – which will remain a magnet for as long as the current is flowing in the circuit. This gives us a magnet we can switch on and off, called an **electromagnet**.
- The nail acts as an iron core inside the coil of wire – this makes the magnetic field stronger than if it was just the field around the wire itself.

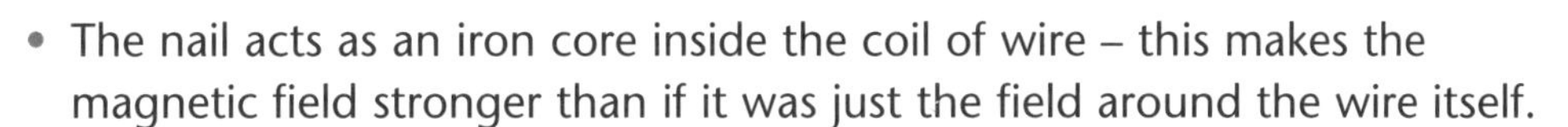

- You can also make the magnetic field stronger by having more turns on the coil or by having a larger current.
- Electromagnets differ from ordinary magnets because the magnet can be turned on and off by switching the current on and off.

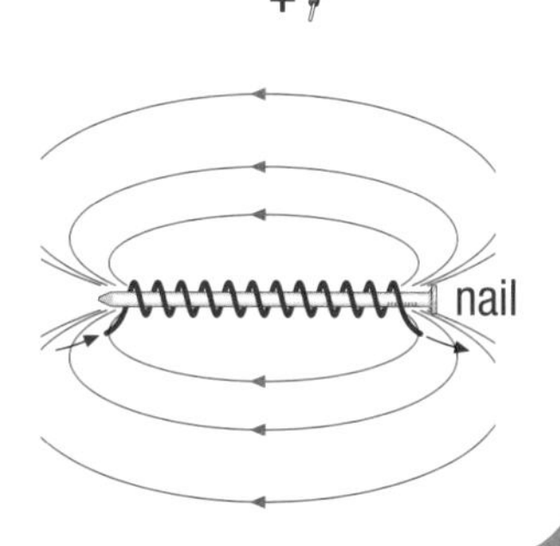

levels 5-6

Using magnets and electromagnets

- Simple applications for magnets include a cat flap that closes.
- Electromagnets are used in many everyday devices such as circuit breakers. They are also used in scrap yards to pick up or drop large quantities of steel from old cars. In hospitals, electromagnets are found in very delicate machines that remove metal splinters from eyes and wounds.
- A more complex application is a door bell that contains an electromagnet which can turn on and off very quickly, making a tiny hammer hit a bell many times a second to keep the bell ringing.

1. What **two** things can you change to make an electromagnet stronger?
2. What is the advantage of an electromagnet compared to an ordinary magnet?
3. Where will the North-seeking pole of a magnet always point to?

Solids, liquids and gases

States of matter

level 3

- All matter exists in one of three states: **solid**, **liquid** or **gas**. These states of matter all have different arrangements of their **particles**.

Particles in solids

level 4

- In a solid, the particles are tightly packed in a regular lattice pattern. The bonds between the particles are strong and individual particles do not have enough energy to escape from the lattice. This is why solids have fixed shapes. The particles are not still – they vibrate within the lattice – but they cannot break free from it.
- It is important to draw the particles in a solid in this kind of regular, even, lattice structure.

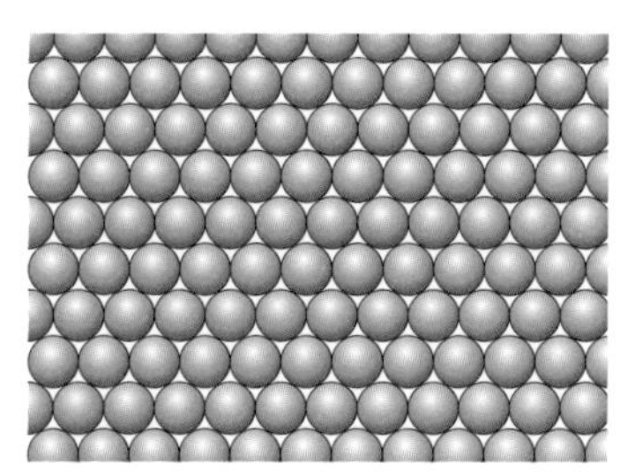

Solid

Particles in liquids

level 4

- When a substance is given more energy – usually by heating – the particles start to vibrate more and more until they have so much energy that the lattice structure starts to break down and the particles begin to slide over each other. The substance has then become a liquid. Liquids can flow and they take up the shape of their container because the particles are able to move more freely while still remaining in contact with one another. Liquids cannot be squashed.

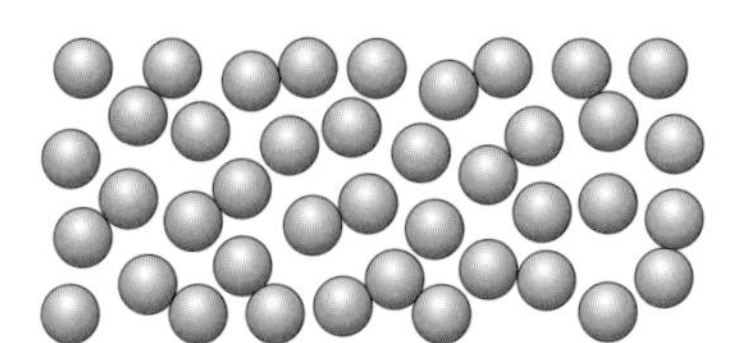

Liquid

- Particles in a liquid are the most difficult to draw. You must have two-thirds of the particles in contact with each other – but it must not be a regular pattern.

Particles in gases

level 4

- If even more thermal energy is put into the substance, then the particles move more and more freely until they can escape the bonds between them completely. The substance has then become a gas. Gases spread out to fill whatever space is available because the particles are free to move right away from each other. Gases can be squashed because there is so much space between the particles.
- Gases are the easiest to draw as there are only ever a few particles – well spaced out.

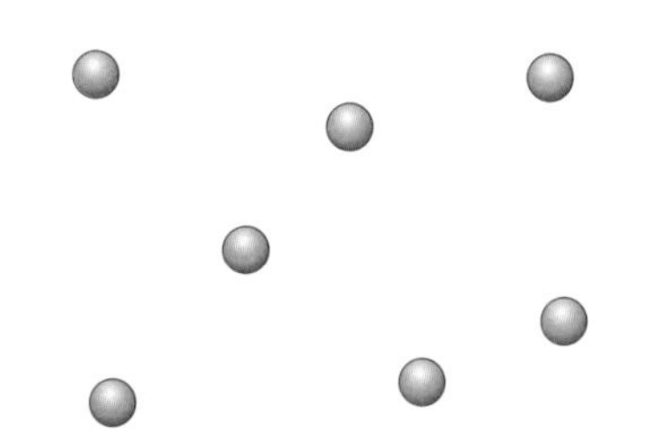

Gas

Changes of state

- A substance can change from a solid to a liquid, from a liquid to a gas, from a gas to a liquid or from a liquid to a solid. We describe this as a **change of state**.
- These changes are **physical changes**, NOT **chemical changes**. The substance remains the same – it is just in a different form – and these changes can be easily reversed.
- If you think of a beaker of hot water, it does let off steam but not all the water changes to steam at once. This is because the particles do not all have the same amount of energy. The most energetic particles will escape the liquid water as steam and the less energetic ones will remain as water.

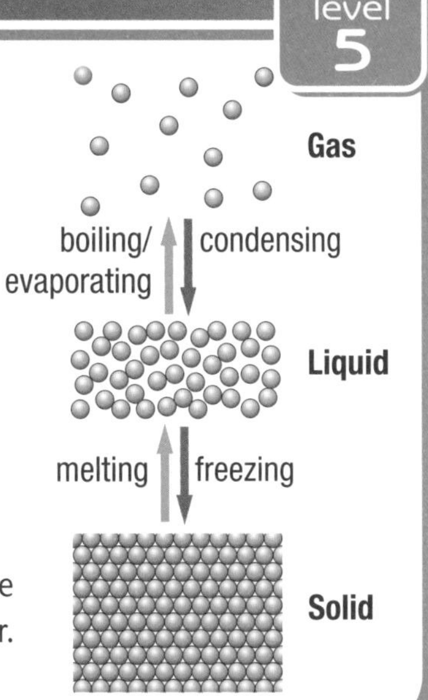

Top Tip!

Remember that water is a liquid, but water vapour is a gas.

More than one state

- Elements and compounds can all exist in more than one state of matter. Very cold oxygen and nitrogen will be liquids, as will very hot metals.
- The most familiar state for any substance will be its state at normal temperatures – oxygen as a gas and metals as solids – but they can exist in other forms.

Did You Know?

The temperature –273 °C is called Absolute Zero. It is thought that at this temperature everything would be solid – but no-one knows how to get any substance to this temperature.

level 4

Expansion and contraction

- When solids and liquids get hotter, they **expand** because the particles move further apart. When they cool, they **contract** because the particles become closer together. This is why, for example, railway lines are built in sections with small gaps in between to allow the rails to expand in hot weather.
- When solids and liquids expand, the particles move more and spread out more – the particles themselves do NOT get bigger. It is like living things growing because the cells divide and make more cells rather than the cells getting bigger.

1. What is the correct name for each of these changes of state:
   a from solid to liquid; **b** from gas to liquid?
2. How would you describe the arrangement of particles in a solid?
3. What are the particles in a liquid able to do that means a liquid can take on the shape of any container?

PARTICLES Solubility and separation

level 4

Solubility

- When a solid dissolves in a liquid, we say that the solid is **soluble** in that liquid. The liquid is called a **solvent** and the solid a **solute**. Together they make up a **solution**.
- Many solids will dissolve in some liquids but not in others. When a solid dissolves in a liquid, there is no chemical change – the solid substance is still there. It is may be less easy to see, but it has not gone away. You can demonstrate this quite easily:

Solvent  **Solute**  **Solution** 

if you take 100 g of solvent ... and add 10 g of solute ... you will have 110 g of solution.

The solid particles of the solute have spread out among the particles of the liquid solvent so you can no longer see them, but the increased mass of the liquid shows that they are still there.

- If a solid will not dissolve in a particular solvent, we say it is **insoluble** in that solvent.
- If you put sand into water, you can still see the sand because it is insoluble. If you put salt into water you cannot see it because it dissolves.

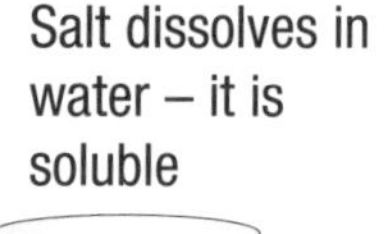

Top Tip!

Don't confuse the term solvent (the liquid) with the term solute (the solid), both of which make up the solution.

Particles and solubility

- If a solid dissolves, it is because the particles are small enough to fit into the spaces between the particles of the liquid.
- If a solid does not dissolve, it is because the particles are too big. More solid can dissolve in a hot liquid as there is more space between the liquid particles when they have more thermal energy.
- This is important when we start to think about how to separate the dissolved particles from the liquid solvent.

Spot Check

1. Fill in the blanks in the passage by putting in the correct word from the list. You may need to use a word more than once.
   solution insoluble solute solvent dissolve
   To make a ...., take some .... and dissolve it in a .... If your solute will not .... in your solvent then it is .... Every solute will not dissolve in every ....
2. What method or methods of separation would you use to get sand and salt from water containing both solids?

Did You Know?

Distillation has been around for a very long time – its first recorded use was the distillation of wine in the City of Bath in the 1200s.

level 4

Separating techniques

One difference between a mixture and a compound (see page 59) is that it is quite easy to get back the substances that went into a mixture and very difficult to get back the substances that reacted together to make a compound. There are several techniques that can be used to separate mixtures.

level 4

Filtration

- If you have a mixture of a liquid and a solid that has not dissolved, you can **separate** the liquid and solid by **filtering**.
- The mixture is poured through filter paper, which is full of tiny holes. The large, solid particles are too big to go through the holes and are left in the paper, but the smaller particles of liquid pass through – so the solid and liquid are separated.
- Filtering works for sand and water, but not salt and water because the tiny particles of dissolved salt go through the paper with the water.

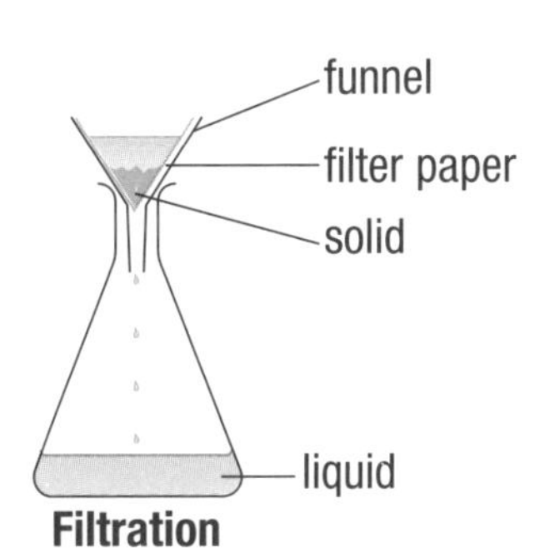

Filtration

levels 4-5

Evaporation and distillation

- Water boils at 100 °C and becomes steam. To get pure water from sea water, you need to cool and condense the steam or water vapour produced when water boils, so that it turns back into liquid water. This process is known as **distillation**.

Evaporation

- When you heat a solution of salt and water, all the water will evaporate leaving the salt behind. **Evaporation** is a good separation technique if you only want to get the salt from the solution and not the water.

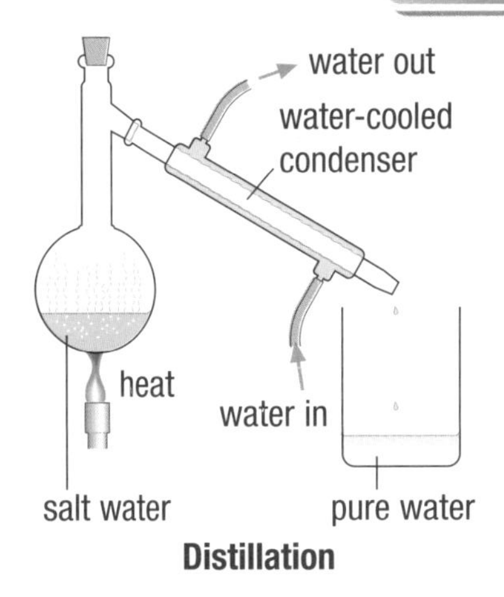

Distillation

levels 4-5

Chromatography

- When more than one solute is dissolved in a solvent, they can be separated using **chromatography**. In chromatography, the solutes, which all have different sized particles, move at different speeds across filter paper.
- Coloured inks are solutions made from several solutes with different colours (or pigments) dissolved in a single solvent. The solutes have different sized particles and can be separated by chromatography, creating a pattern of coloured dots or lines on the filter paper.

Elements, compounds and mixtures

level 5

Elements

- An **element** is a pure substance. All the atoms in an element are of the same kind. For example, oxygen is made up of just oxygen **molecules**, and copper is made up of just copper **atoms**.

level 6

The Periodic Table

- All the known elements are listed in the **Periodic Table**. The elements are listed in order of how big their atoms are, and elements that are like each other are grouped together.
- All the metals are on the left-hand side and all the non-metals are on the right-hand side.
- The more reactive a metal is, the nearer it is to the left-hand side (see page 65 for more on reactivity). The elements in the last column on the right hardly react at all.
- In the Periodic Table, elements are represented by symbols. There are about 100 symbols. You do not need to know the whole of the Periodic Table, but you should know the names and symbols of the elements given below.

Element	Symbol	Element	Symbol	Element	Symbol
Hydrogen	H	Magnesium	Mg	Copper	Cu
Helium	He	Aluminium	Al	Zinc	Zn
Lithium	Li	Phosphorus	P	Lead	Pb
Neon	Ne	Sulphur	S	Gold	Au
Boron	B	Chlorine	Cl	Silver	Ag
Carbon	C	Argon	Ar	Sodium	Na
Nitrogen	N	Potassium	K	Nickel	Ni
Oxygen	O	Calcium	Ca	Iron	Fe

level 5

Combining elements

- Everything is made up of different combinations of the 100 or so elements. It is a bit like all the words in all the books in all the libraries being made up from the 26 letters of the alphabet!
- Just as letters go together differently to make up different words, so elements go together differently to make up different substances that we call compounds.
- Some molecules of some elements have only a single atom. But some elements, including hydrogen and oxygen, are made up of molecules which have two atoms each: H_2 and O_2.

Did You Know?

The Periodic Table was devised by a Russian scientist Dimitri Mendeleev in the late 1800s. He left spaces for elements that hadn't yet been discovered.

level 5

Compounds

- When two or more elements react together in a **chemical reaction**, a compound is formed. A compound is often very different from the elements that react together to make it – and it is very difficult, or impossible, to get the elements back.
- Sodium chloride, the salt that we eat, is a compound. The **formula** for sodium chloride is NaCl.
  NaCl is a white solid better known as salt. It is made from sodium, which is a very reactive, soft, grey metal that reacts violently with water and chlorine, which is a green poisonous gas that is used as a disinfectant in swimming pools.

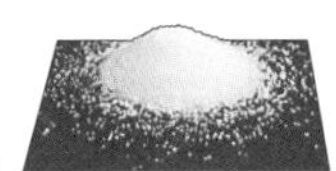

A mixture is a combination of substances that have NOT reacted with each other. A compound is a substance that is made when a chemical reaction takes place.

level 5

Mixtures

- Two or more elements, or compounds, can be mixed together in a way that does not include any chemical reaction. This is a **mixture**.
- The elements or compounds in the mixture do not change their nature and can be easily separated.

level 5

Recognising elements, compounds and mixtures

- In an element all the atoms will be the same, even if they exist as molecules.
- In a **compound** the molecules will all be the same, but will be made up of different atoms joined together.
- In a mixture there can be any variety at all, but the atoms will certainly not all be the same.

This could be oxygen.

Element

This could be magnesium oxide.

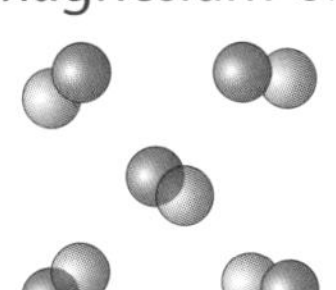

Compound

This could be any mixture – for example air.

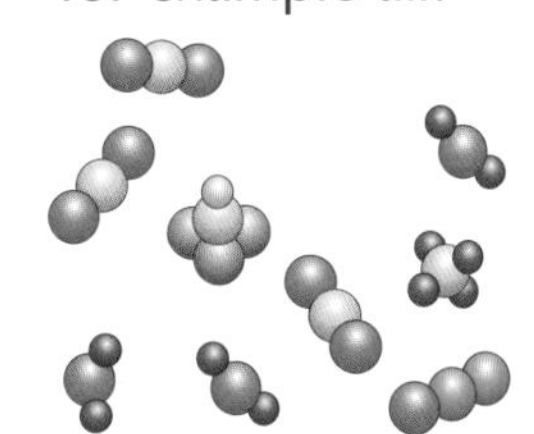

Mixture

Spot Check

1. Where on the Periodic Table would you find the elements that are metals?
2. The chemical formula for ethanol is C_2H_5OH. Which elements react together to make ethanol?
3. If a molecule of magnesium oxide has the formula MgO, how many atoms does it contain and of which elements?

Chemical reactions

levels 4-5

Making new substances

- **Chemical reactions** between elements and compounds are how new substances are made. Chemical reactions can be complicated industrial processes or they can be everyday actions such as cooking food and burning fuels.
- Chemical changes are different from physical changes of state in some important ways:
  - In a chemical change, new substances are formed.
  - In a physical change, the substance stays the same – it just exists in a different form.
  - Chemical changes are very hard – sometimes impossible – to reverse. You cannot 'uncook' a cake or 'unburn' a lump of coal.
  - Physical changes of state are quite easy to reverse. You can melt ice or condense steam to get the water back.

level 5

Energy changes in chemical reactions

- Chemical reactions always involve an energy change. Sometimes you have to put energy in to make the reaction happen – these are called **endothermic reactions**. In other reactions, such as the burning of fuels, energy is given out – these are called **exothermic reactions**.

level 5

Oxidation

- **Oxidation** describes any reaction where another substance reacts with oxygen. For example:

  Magnesium + oxygen → magnesium oxide

- Many oxidation reactions occur in our everyday lives. If you leave an apple after you have taken a bite from it, the inside of the apple will go brown – this is oxidation. When iron rusts, we are seeing the formation of iron oxide – another oxidation reaction – as some metals can combine with the oxygen present in the air.

Top Tip!

When something burns in air, it is the oxygen present in the air that is used in the reaction.

level 5

Combustion

- **Combustion** describes the reaction that happens when any substance burns in oxygen. Combustion reactions such as the burning of fossil fuels give out a lot of heat.
- Not all oxidation reactions are examples of combustion, but burning always needs oxygen. This is why oxygen is one of the sides of the fire triangle which shows the three things that a fire needs to keep burning.

Chemical equations

levels 5-6

- **Chemical equations** are short-hand ways of describing chemical reactions.
- They can either be written as word equations or using **symbols** and **formulae**.

Examples of chemical equations

level 7

- The oxidation reaction that happens when magnesium burns in air can be written as:

  Magnesium + oxygen → magnesium oxide

  or as:

  $$2Mg + O_2 \rightarrow 2MgO$$

  As oxygen molecules have two atoms (O_2) and magnesium atoms exist on their own, we need two magnesium atoms, one for each oxygen atom, so that is why we write 2Mg.

- The same sort of equations describe the making of salt:

  Sodium + chlorine → sodium chloride

  $$2Na + Cl_2 \rightarrow 2NaCl$$

Did You Know?

The flame in a Bunsen Burner gets as hot as 1500 °C – so take care in school science lessons!

Photosynthesis as an equation

levels 4-5

- The word equation for photosynthesis is more complicated:

  Water + carbon dioxide —(chlorophyll / sunlight)→ glucose + oxygen

- The **reactants** are on the left of the arrow and the **products** on the right of the arrow. The things that have to be there for the reaction to happen, but do not actually react, are written above and below the arrow.

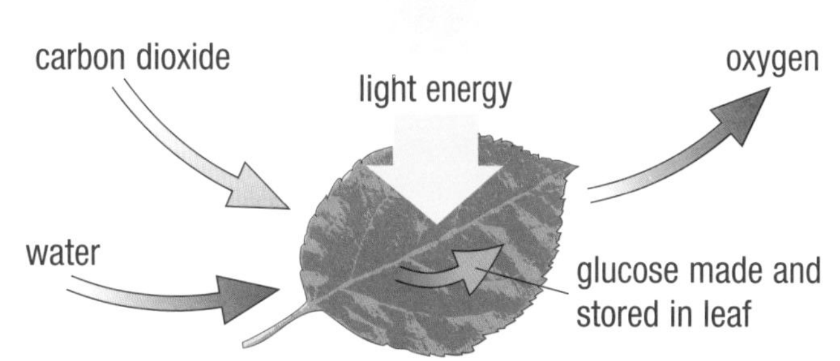

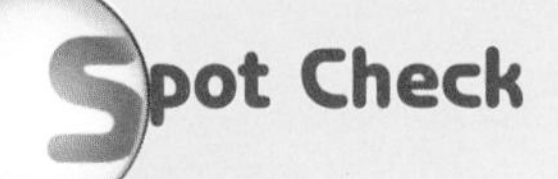

1. What is the correct name for a chemical reaction that gives out heat?
2. Write the equation for the oxidation of magnesium in words and symbols.
3. What name is given to the oxidation reaction that releases energy from fuels?

PARTICLES Acids and alkalis

level 4

Acid, alkali or neutral?

- When anything is dissolved in water, the solution it forms is either an **acid**, an **alkali** or **neutral**.

ACIDS: Everyday acids include fruit juices, vinegar and a lot of fizzy drinks!

ALKALIS: Soaps, toothpaste and detergents are generally alkaline solutions.

NEUTRAL: Pure water is neutral – a neutral solution is neither acidic nor alkaline.

levels 4-5

The pH scale

- The **pH** scale is a set of numbers from 1 to 14 that tells us not only whether a solution is acid or alkali but how strong it is:
  - very strong acids have low pH values from 1 to 3 or 4, weak acids are around 5 or 6
  - neutral solutions are 7
  - A pH above 7 means the solution is alkaline, the higher the number the more alkaline it is.

levels 4-5

Detecting acids and alkalis

- Looking at a solution does not tell us if it is an acid or an alkali. We need to use **indicators** to tell us the pH of the solution. Indicators are substances that turn different colours in acidic and alkaline solutions. A very simple indicator is **litmus** which turns red in acid and blue in alkali.
- **Universal indicator** has a range of colours to show the change in the pH of the solution in more detail.

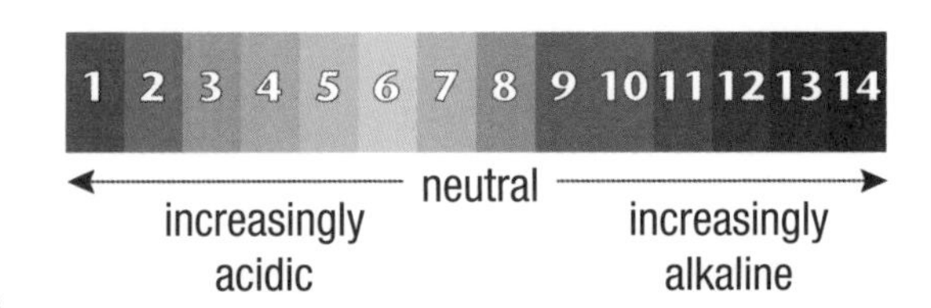

Top Tip!

It is easy to think of acids being dangerous – but many everyday foodstuffs are acid and do us no harm at all.

level 5

Neutralisation

- When you mix an acid with an alkali, the resulting solution will be closer to neutral than either of the original ones. If you mix them in exactly the right amounts you will end up with a solution with a pH of exactly 7. Medicines to treat indigestion are alkalis designed to neutralise the excess stomach acid which causes the pain, but not all the acid as stomach acid is essential for digestion.

levels 5-6

Neutralisation reactions to produce a salt

- Neutralisation produces a **salt** and water. In this sense 'salt' means a group of compounds that are formed when acids and alkalis react: **Acid + alkali → salt + water**
- The common salt that we eat, sodium chloride, is just one of this group of chemicals:

  Hydrochloric acid + sodium hydroxide → sodium chloride + water

  Other examples are: **Sulphuric acid + potassium hydroxide → potassium sulphate + water**
  Nitric acid + calcium hydroxide → calcium nitrate + water
- Sodium chloride, potassium sulphate and calcium nitrate are all salts.
- The type of acid is what controls the kind of salt produced:
  - hydrochloric acid gives **chloride** salts
  - sulphuric acid gives **sulphate** salts
  - nitric acid gives **nitrate** salts.

Did You Know?

Some fizzy drinks have the same pH as sulphuric acid – remember to brush your teeth!

level 6

Acids and metals

- Acids react with metals in a very similar way: **Acid + metal → salt + hydrogen**
- A lighted splint held close to the top of a test tube full of hydrogen will produce a 'pop'.

levels 6-7

Acid rain

- Burning fossil fuels releases gases such as sulphur dioxide into the air. These gases dissolve and make the rain acidic. Acid rain falling into lakes and rivers changes the pH of the water and so affects the plants and animals that live there.
- Acid rain can also effect the pH of soil so that plants that have lived in a particular habitat can no longer survive.
- Acid rain falling onto buildings can dissolve the stone and over a long period cause severe damage to the building.

Trees affected by acid rain

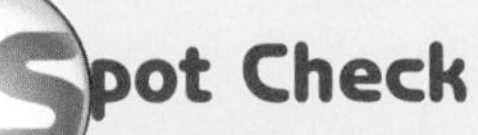

1. If a solution turned Universal Indicator red, what type of substance would you need to add to turn the indicator green? Explain your answer.
2. Complete these word equations for the reactions between an acid and an alkali and an acid and a metal:

   a Acid + alkali → **b** Hydrochloric acid + sodium hydroxide →
   c Acid + metal → **d** Zinc + sulphuric acid →

Metals and non-metals

level 4

Properties of metals and non-metals

- The elements in the Periodic Table are divided into **metals** and **non-metals**. Metals and non-metals have distinct **properties** (with some exceptions).

Properties of metals (such as copper, iron, zinc)	Properties of non-metals (such as wood, stone, glass)
All metals are good conductors of heat	Non-metals are generally poor conductors of heat
All metals conduct electricity	Non-metals do not conduct electricity – except graphite
Metals are all solids at normal temperature – except mercury which is a liquid	Non-metals can be solids, liquids or gases at normal temperature
Metals generally have a shiny appearance and are often grey	Non-metals have a wide variety of appearances

Top Tip!

Metals feel cold to the touch because they conduct the heat away from your skin. Non-metals feel warm because they are poor conductors of heat.

level 5

Properties and use

- The properties of metals and non-metals are linked to the things we use them for. For example, saucepans are made of metal because they **conduct** heat from the cooker to the food. But you stir the contents with a wooden spoon because it does not conduct heat – so your fingers will not get burnt.

- You need to know the properties of metals and how they are linked to their uses. However, not all metals are the same and cannot be put to the same uses. For example, you cannot make water pipes out of iron because it rusts – water pipes are normally made from copper which does not react with water.

level 5

Corrosion

- Metals react with oxygen to produce metal oxides:
  Metal + oxygen → metal oxide
- This reaction is called **corrosion**. With some metals, like magnesium, aluminium and lead, this corrosion produces a thin layer of the metal oxide which makes the surface look dull instead of shiny. Corrosion is usually quite a slow reaction – some metals like gold and platinum, which are not very reactive, do not corrode at all.
- The corrosion of iron is called **rusting**. Rusting needs oxygen and water:
  Iron + oxygen + water → hydrated iron oxide (rust)

level 6

Reactivity series of metals

- Some metals react far more readily than others with water, steam and acids:

  Metal + water → metal hydroxide + hydrogen

  Metal + steam → metal oxide + hydrogen

  Metal + acid → salt + hydrogen

- The metals at the top of the list on the right react with cold water – at the very top they react quite violently. The metals in the middle do not react with cold water but they do react with steam. The ones at the bottom don't even react with steam.
- Metals as far down as copper and mercury also react with dilute acids.
- Silver also reacts with oxygen but gold and platinum do not even do that. This is why silver tarnishes but gold jewellery does not.
- The nearer they are to the top the more rapidly they react.

Potassium
Sodium
Calcium
Magnesium
Aluminium
Zinc
Iron
Nickel
Tin
Lead
Copper
Mercury
Gold
Platinum

Did You Know?

Gold jewellery found in the Pharaohs' tombs in Egypt was completely untarnished after thousands of years because gold does not react.

level 7

Displacement reactions

- If an iron nail is placed in a solution of copper sulphate, the nail becomes coated with copper. This is because the less reactive copper is displaced (or pushed out) of the copper sulphate by the more reactive iron. The word equation for this reaction is:

  Copper sulphate + iron → iron sulphate + copper

- You can always predict when a metal will displace another from a salt by looking at the reactivity series. More reactive metals always displace less reactive ones – never the other way around. If you put silver into a solution of copper sulphate there would be no reaction and the silver would continue to look just the same.

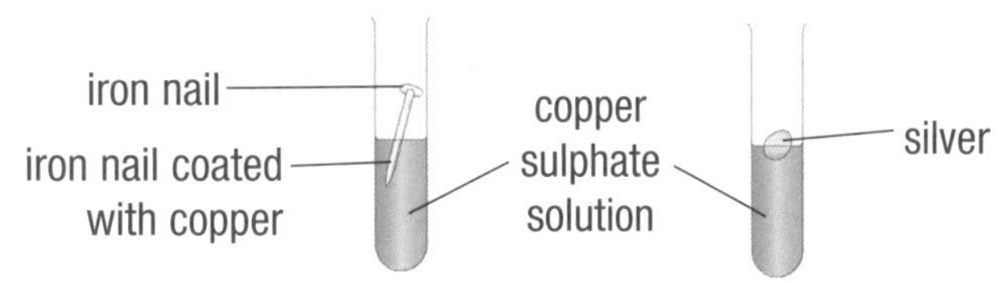

- This would be the same for any other metal – those above copper would displace the copper from the copper sulphate solution and those below would not.

1. What **two** things does iron have to be protected from to stop it rusting?
2. What is the word equation for the reaction of a metal with steam?
3. Which of these metals would displace lead from lead nitrate?
   magnesium **silver** **copper** **iron**

Rocks and the rock cycle

level 3

Types of rock

- There are three main groups of rock:

igneous **sedimentary** **metamorphic**

levels 4-5

Igneous rocks

Granite rocks

- The inside of the Earth is very hot and some of the rock is molten. The molten rock is called **magma**. When magma cools it forms crystalline rocks called igneous rocks.
- If the magma escapes onto the Earth's surface through a volcano and then cools quickly, it forms rocks with small crystals. If it cools more slowly underground, then the crystals in the rock will be bigger.
- Granite is an example of an igneous rock with large crystals. Basalt is an igneous rock with small crystals.

Top Tip!

At level 7 you will need to be able to apply your wider knowledge of chemical reactions to the causes and effects of geological changes.

level 5

Sedimentary rocks

- When rainwater gets into rocks and then freezes and thaws, it can cause pieces of rock to break off. This is called **weathering**.

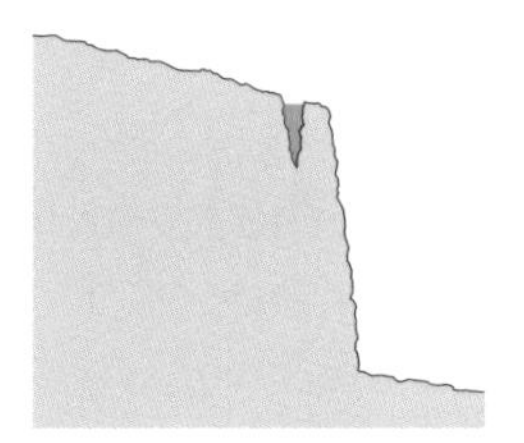

Rainwater collects in a crack.

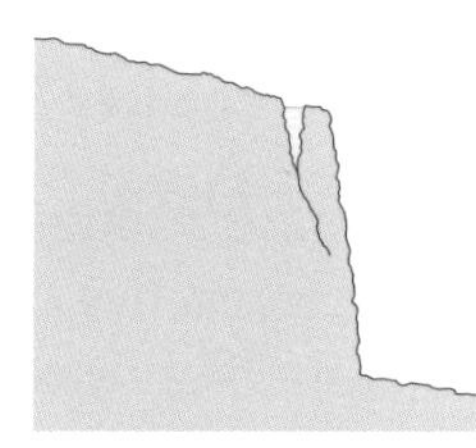

The temperature falls below 0 °C. The water freezes and expands, making the crack bigger.

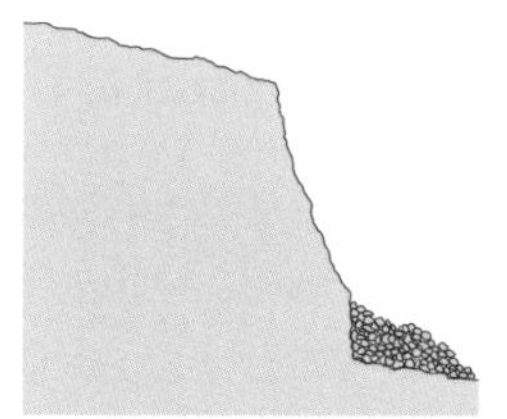

Eventually, after repeated freezing and thawing, the rock breaks off.

- The small pieces of rock are moved by rivers and glaciers and are gradually worn smooth. These pieces of different-sized rock sink to the bottom of the sea and over many, many years the sediment is stuck together and the water squeezed out, forming sedimentary rocks.
- Sandstone, chalk and limestone are examples of sedimentary rocks. Sometimes the remains of animals and plants are caught among the sediment and are preserved and compressed to form **fossils**.

Chalk rocks

- Rocks can also be broken down by erosion which can be the physical action of wind or water over many years wearing away the rock, or chemical erosion where rocks are worn away by chemical reactions with substances such as acid rain.

levels 4-5

Metamorphic rocks

- The word metamorphic means 'changed', so rocks that have changed are called metamorphic rocks.
- Layers of sedimentary rocks are changed by underground heat or underground weight, or both. They have a layered or streaky appearance.

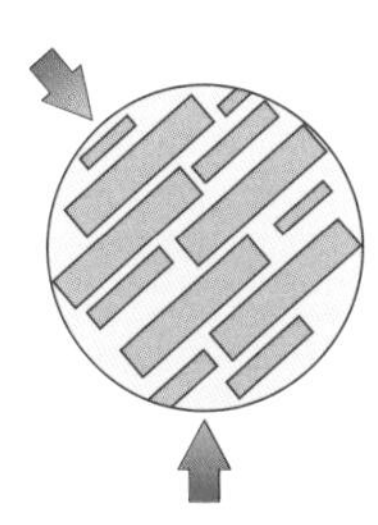

The fragments that make up sedimentary rocks are deposited in layers.

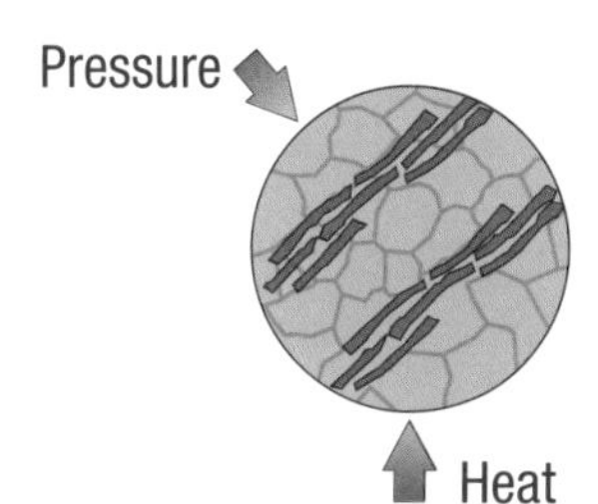

The rock is heated and squashed (compressed) and new layers start to form.

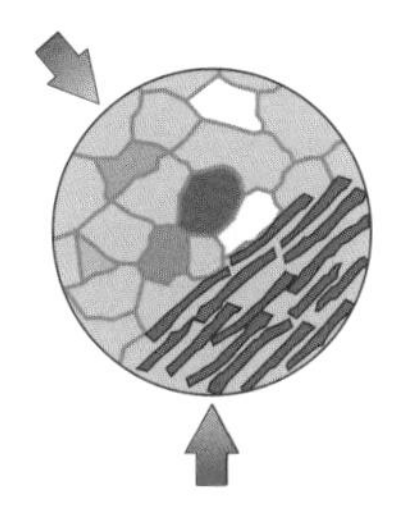

This heat and pressure cause new minerals to form, often crystals in layers.

- When limestone is heated and compressed it forms marble. Sandstone changes into a much harder rock called quartzite.

Did You Know?

One person in every 10 in the world lives within range of an active volcano.

level 5

The rock cycle

- The surface of the Earth is always changing and the three groups of rocks recycle from one kind to another.

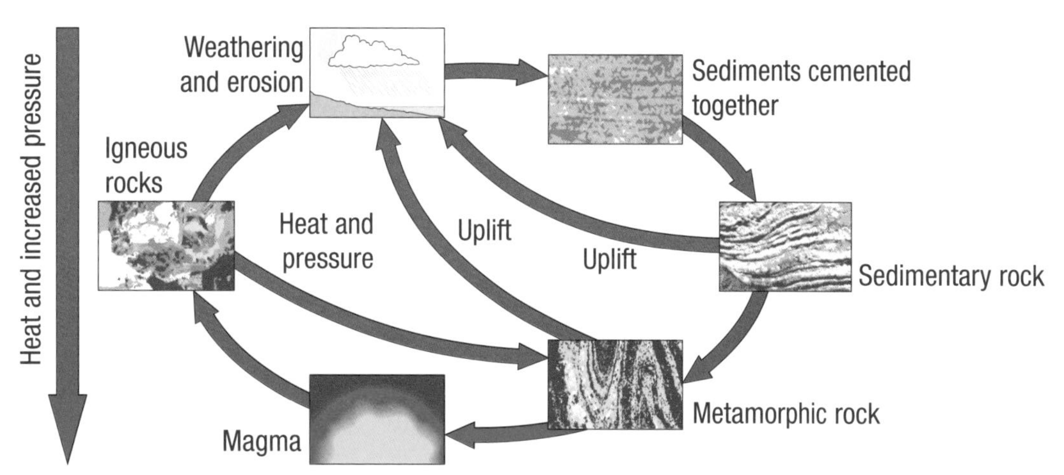

1. How can we tell by looking at an igneous rock how quickly the magma cooled to form this rock?
2. In what sort of rock do we find fossils and why?
3. What **two** physical quantities act on sedimentary rocks to produce metamorphic rocks?

INTERDEPENDENCE Food chains and food webs

level 3

Obtaining energy

- All living things need energy to enable them to carry out life processes.
- Plants are able to trap the energy of sunlight and make their own food.
- Animals have to eat plants or other animals to obtain the energy they need.

levels 3-4

Food chains

- A green plant is always at the start of every food chain. These plants are called **producers**.
- Animals that eat plants are called **primary consumers**. They are **herbivores**.
- Animals that eat the primary consumers are called **secondary consumers**. They are **carnivores**.
- Then **tertiary consumers** may eat the secondary consumers. These are also carnivores.

PRODUCER → PRIMARY CONSUMER → SECONDARY CONSUMER → TERTIARY CONSUMER

GREEN PLANT → HERBIVORE → CARNIVORE → CARNIVORE

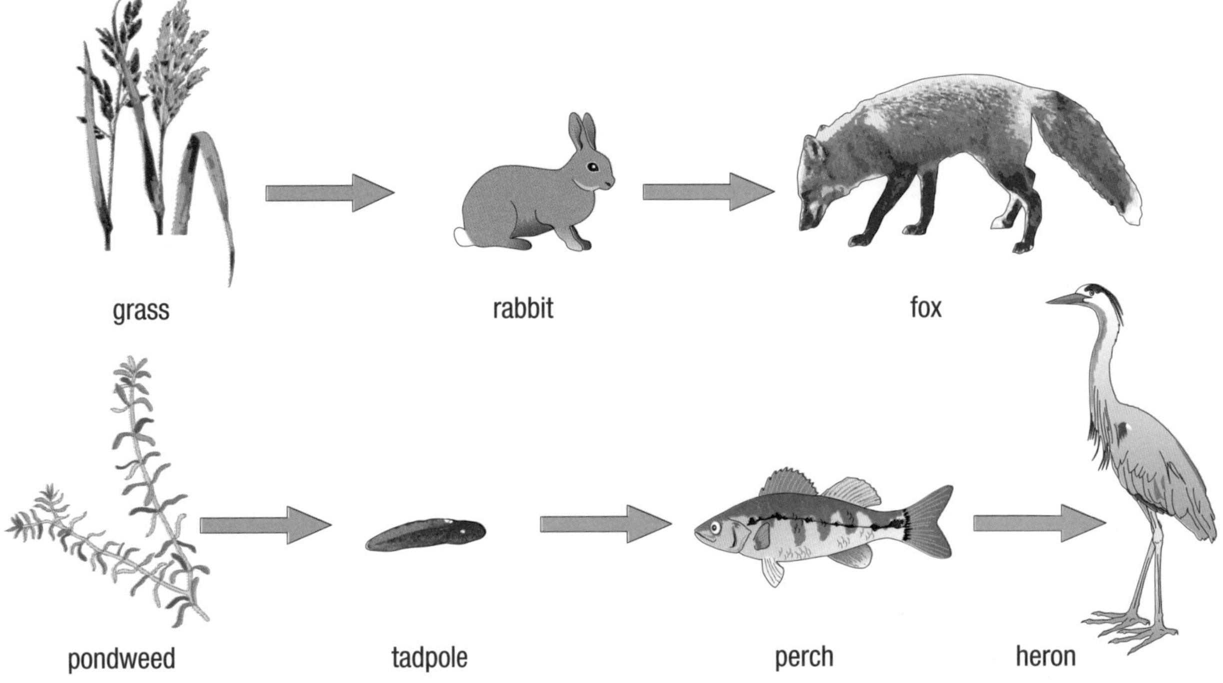

Top Tip!

The arrow in a food chain always goes from food to the creature that eats it. It shows the way in which the energy moves through the food chain.

Did You Know?

Owls can fly almost silently so that their prey will not hear them coming – but they have very sensitive ears themselves to help them find their prey at night.

level 4

Food webs

- In reality, many food chains are not as simple as this. For example, most humans are neither only herbivores nor only carnivores – they eat both plants and animals. They are called **omnivores**. This is less usual among other animals, though.
- In any habitat there will be several food chains connected together to make **food webs**.
- A woodland food web might look like this:

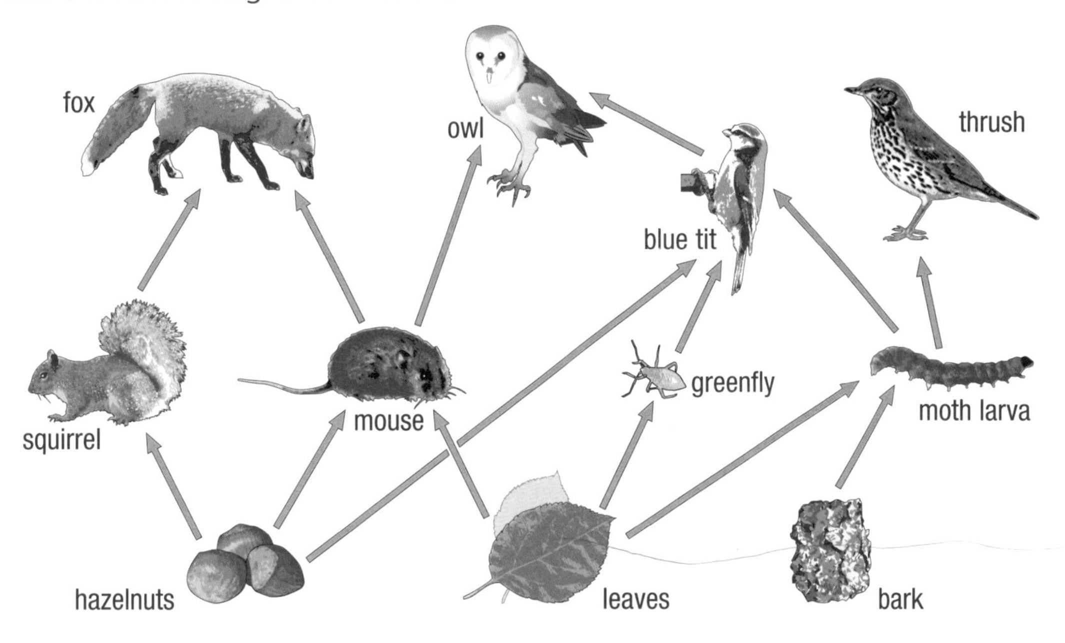

level 5

Pyramid of numbers

- There are not the same numbers of individuals at each stage – one blue tit will eat a lot of greenfly and one owl will eat several blue tits. We can show this in a **pyramid of numbers**.

owl
blue tit
greenfly
leaves

1. What kind of organism is a producer and why is there always one of these at the start of every food chain?
2. Draw a pyramid of numbers for this food chain:
   pondweed → tadpole → perch → heron
3. In a food chain, how would you describe the diet of the secondary consumer?

Energy transfers in food chains and food webs

level 6

Passing energy on

- Food chains and webs always have a green plant at the start because plants can trap the energy from sunlight. This energy is then passed up the food chain to the other organisms.
- Only a small amount of the energy taken in can be passed on up the food chain, the part of the energy the organism uses for growth. Think of the first food chain on page 68:

  The fox is a **predator** that eats the rabbit that is its **prey**.

  GRASS → RABBIT → FOX

- One rabbit will feed on a lot more than one grass plant and a fox will expect to eat more than just one rabbit in its life! This gives us a pyramid of numbers, as we saw before:

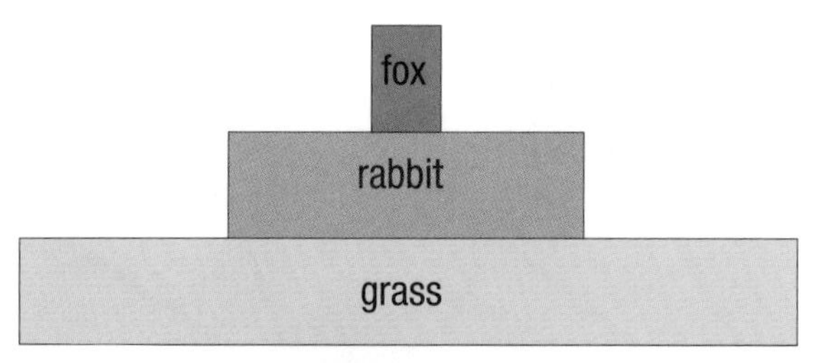

level 6

Using and losing energy

- If all the energy in all the grass were transferred to all the rabbits and all the energy in all the rabbits were transferred to all the foxes, then there would not be such a big change in the number of individuals between each level. However, grass plants don't just use the energy from the Sun to grow – they also use some of it for other life processes.
- The rabbit is even more 'wasteful' of energy. Some of the energy the rabbit takes in from the grass is transferred to the surroundings as heat, so that the rabbit can maintain its body temperature. Some of the energy is used as movement energy as the rabbit moves around.

Top Tip!

Everything you know about energy transfers applies to energy transfers in food chains and webs – particularly the fact that energy is always lost as heat at each stage of the transfer.

Did You Know?

Humans are the most numerous mammals on Earth, closely followed by rats.

level 6

Where does the energy go?

- The percentages in the picture below are approximate and vary between species, but the proportion of energy passed up the food chain to the next **trophic level** is never more than 10%. This is why food chains are very short. There is just too much energy lost at each stage for there to be more than one or two, or at the most three, **consumers** after the green plant **producer**.

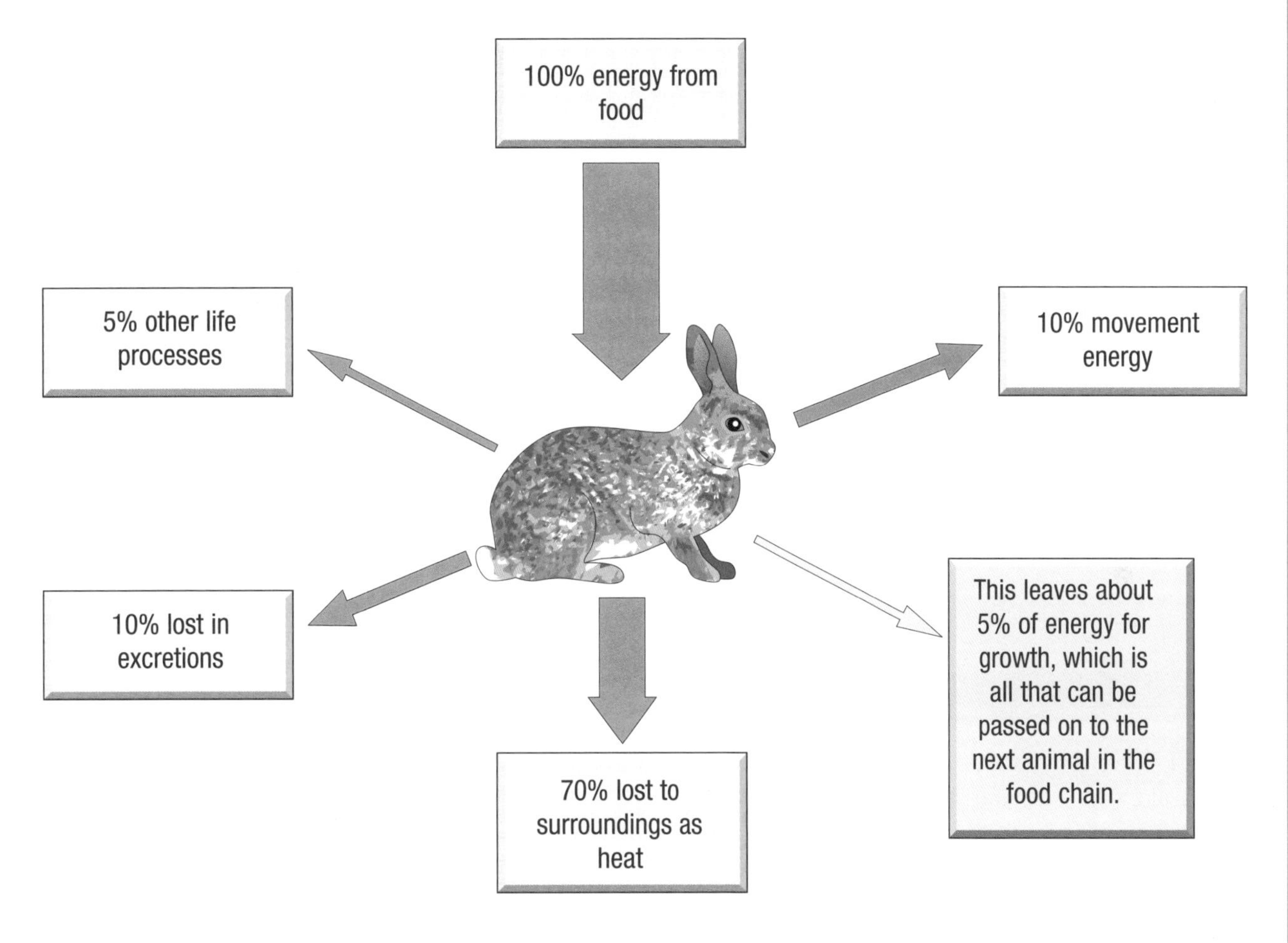

- Food webs can be spread out and be very complicated but there are never more than three consumers after the producer.

1. What names do we give to an animal that eats another animal, and to the animal that gets eaten?
2. Why is all the energy that a plant traps from sunlight not passed all the way along the food chain?
3. Approximately how much energy that an animal takes in is transferred to the next creature in the food chain?

Variation and inheritance

level 4

Variation within species

- Members of the same species have a lot of things in common – but they have differences as well. Humans can have different eye, skin and hair colour. These differences are called **variations**.

levels 4-5

Inherited and environmental factors

- Most of our physical **characteristics** are **inherited** from our parents; others are affected by the **environment** in which we live. The colour of your eyes and the natural colour of your hair will be inherited, but if you choose to change the colour of your hair or put on weight, then these are environmental factors.
- Some things such as height are a mixture of inheritance and environment. We all inherit a maximum possible height but we will only reach that if we have the right diet when we are growing.

level 5

Inheritance

- A sperm and an egg cell each contain half of the DNA that is found in ordinary cells. When the sperm and egg cell join, these two halves of DNA fuse together to form a person who will get half their characteristics from each parent. Different combinations of the genes from each parent produce individuals with different characteristics.
- Identical twins have the same genes so they have the same inherited characteristics. This is because they are made when one fertilised egg divides in two.
- Other brothers and sisters often look quite different as it depends upon which characteristics they inherit from which parent. This includes non-identical twins who are made when two separate eggs are fertilised and implanted in the mother's uterus at the same time.

Remember, our parents pass our inherited characteristics down to us, but some of our characteristics can be changed by our environment.

Did You Know?

Humans share 98% of their genetic material with chimpanzees and 50% with bananas.

level 4

Species characteristics

- Other **species** also show great variation within their members.
- Although dogs are members of the same species, they can have very different characteristics, as shown in this picture.

level 6

Selective breeding

- Members of the same species are those animals or plants that can breed together and produce offspring that can then breed in their turn.
- **Selective breeding** aims to produce more individuals with a characteristic that is thought to be valuable.
- Cattle breeders will select animals with the best milk production and maybe a natural resistance to some diseases in the hope that their offspring will have the same good qualities. Plant breeders go to great efforts to produce a plant with a flower of a particular colour, size or shape by carefully selecting parent plants with the desired characteristics.

level 5

Variation between species

- Usually there are great differences between members of different species. Some species, however, have things in common, for example horses and donkeys or rats and mice, but they are still different species and there will be variations between them.
- These are all woodlice and may look very similar but they are actually different species and have variations.
- It is important to remember that if members of different species try to breed together, the offspring, if any, will not be able to breed.

Spot Check

1. Which of these are inherited characteristics in humans?
   eye colour   natural hair colour   height   weight
   ability to speak a foreign language
2. If two members of different species breed together, what will be the drawback with the offspring?
3. What characteristics might a farmer look for in selectively breeding a flock of sheep?

Adaptation

level 3

Fitting into the habitat

- For things to live successfully in their **habitat** they often need to be adapted to their surroundings.
- This **adaptation** has happened over many, many generations because the individuals that had the most useful characteristics were most able to survive and breed and so pass on these useful characteristics.

level 4

Adaptation in animals

Here are two examples of animals that have become adapted to their environments.

- Moles have the characteristics that they need to live underground:
  - a good sense of smell to detect their food
  - large front claws to dig through the earth
  - a streamlined shape to move efficiently through the tunnel created.

- Polar bears have the characteristics needed to live in the cold and snow:
  - a white coat to make them less easy for predators to see
  - thick fur to trap heat and keep them warm
  - sharp teeth for killing prey
  - plenty of fat for insulation and as a food store
  - strong legs for walking and swimming long distances in search of food
  - sharp claws and hair on the soles of their feet to improve their grip on icy surfaces.

Top Tip!

Adaptation is an everyday word, so make sure you understand and can use its scientific meaning correctly.

level 4

Adaptation in plants

- Many plant species have also adapted so that they can live successfully in particular habitats.
- Cactus plants, for example, have adapted to live in the hot, dry conditions of deserts. They have:
  - long roots to find water below the ground
  - swollen stems to store as much water as possible
  - small spines for leaves, which not only reduce water loss because the surface area is small, but also deter animals from eating the plant.

level 4

Daily adaptations

Many species have adapted in ways that allow them to cope with changes in their environment.

- Daily adaptations include:
  - plants with flowers that open in the day to allow pollinating insects in and close at night for protection
  - animals that come out at night rather than during the day – owls hunt at night for other **nocturnal** creatures, using their excellent sense of hearing rather than sight.

Did You Know?

There is a tree-like cactus in desert regions of North America that has flowers 30 cm in diameter.

level 4

Seasonal adaptations

- Seasonal adaptations include:
  - deciduous trees losing their leaves before the winter when there will be little sunlight for photosynthesis
  - animals growing thicker coats in the winter and even changing colour between winter and summer for maximum camouflage
  - animals eating a lot in the autumn when there is plenty of food and then sleeping or **hibernating** in the winter when food supplies are short
  - birds flying away to warmer places in the winter where there is more food. This is called **migration**.

Spot Check

1 If you were going to design a mammal to live in a cold, wet climate with plenty of green plants and very large animals but not many small animals, what adaptations would this creature need to live there successfully?

2 Why do some animals hibernate in the winter? And what do they do to prepare for this?

3 Why do some plants have flowers that open in the day and close up at night?

Classification

Grouping things together

level 4

- When you go into a supermarket you can find the things you want to buy because similar things are all put together. If you want apples, you go to the fruit and vegetable section; if you want ice-cream you go to the freezer. It saves time and makes doing the shopping much easier than if everything were just put anywhere.

Classifying living things

level 4

- Classification of living things is much the same. It works on the basis that different species share common features and so can be put together. It also means that if we observe the characteristics of an unfamiliar organism, we can most probably tell something about it by working out its classification.
- There are over 800 000 species of animals on Earth so they have to be broken down into groups. The first division is **vertebrates** and **invertebrates** – vertebrates have backbones and invertebrates do not.
- The plant kingdom is also divided into groups and there are three other kingdoms: fungi, bacteria and protoctists.

Top Tip!

Not only mammals are animals – birds, fish, frogs, snakes and insects are all animals, and humans are too!

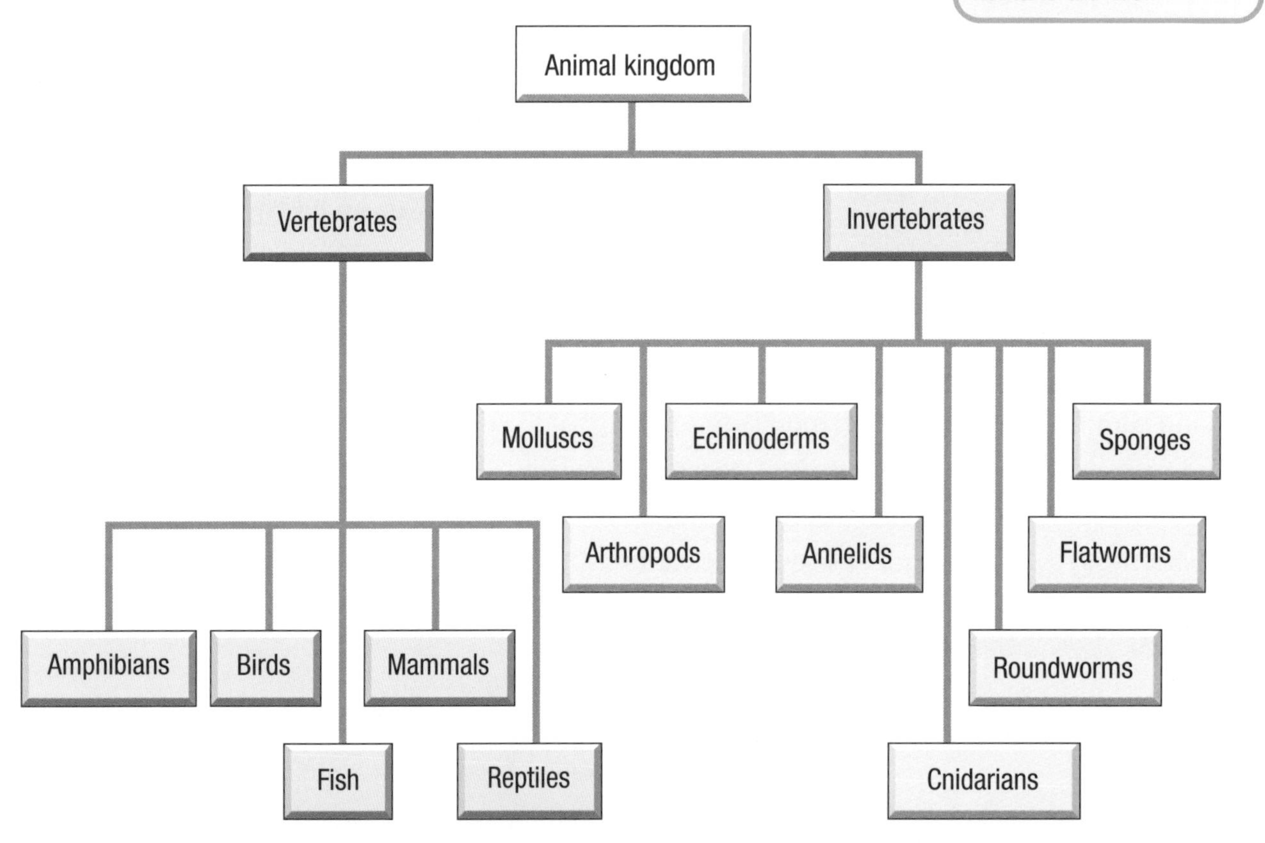

level 5

Characteristics of vertebrates

All vertebrates have a backbone, but their other characteristics are quite different.

- **Amphibians**
  - have moist skins
  - lay jelly-coated eggs in water
  - can live on land or under water.

- **Fish**
  - have a streamlined shape
  - breathe through gills
  - live and lay eggs in water
  - have wet skin with scales.

- **Birds**
  - are warm-blooded
  - breathe through lungs
  - have feathers
  - lay eggs with hard shells.

- **Mammals**
  - are warm-blooded
  - develop the young inside the mother
  - feed their young on milk from the mother
  - breathe through lungs
  - have fur or hair.

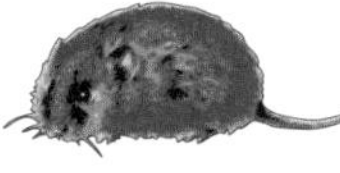

- **Reptiles**
  - have dry scaly skin
  - breathe through lungs
  - lay eggs with a leathery skin on land.

Did You Know?

A large fossil discovered in South Africa is thought to be one of the first vertebrates. The eel-like creature called a Conodont evolved 520 million years ago which is 50 million years before the first fish.

level 5

Characteristics of invertebrates

Arthropods are the biggest group of invertebrates. They can be sub-divided into four groups:

- **Insects**, such as grasshoppers, dragonflies and wasps, have three-sectioned bodies and three pairs of legs.
- **Arachnids**, such as spiders and scorpions, have two parts to their bodies and four pairs of legs.
- **Crustaceans**, like crabs and lobsters, have two pairs of antennae, a chalky shell and up to seven pairs of legs.
- **Myriapods**, such as millipedes and centipedes, have long, thin, segmented bodies with a pair of legs on each segment.

Spot Check

1. In what way are mammals different from other vertebrates?
2. What are the **four** groups of arthropods called?
3. Name **two** characteristics of birds.

INTERDEPENDENCE

Using keys to identify living things

level 4

Types of key

- As there are so many species of living things, it would not be possible for anyone to remember the names of all of them. However, it is possible to find out which species any organism belongs to using something called a **key**.
- Keys may be either a set of questions and answers or a branching diagram. For example, you can use a simple question and answer key to distinguish between some of the 15 000 known species of butterfly. There are 6000 species in the family Nymphalidae alone!

These are pictures of five examples:

A

B

C

D

E

level 4

Using a question key

- Look carefully at the pictures and answer the questions below to name these five butterflies.

1 Does the butterfly have a yellow or white edge to its wings?
- If yes, it is the Camberwell Beauty (**C**).
- If no, go to question 2.

2 Does the butterfly have a noticeable amount of white on its wings but hardly any at the edges?
- If yes, it is the White Admiral (**A**).
- If no, go to question 3.

3 Does the butterfly have blue colouration anywhere on its wings?
- If yes, go to question 4.
- If no, it is the Comma butterfly (**B**).

4 Is the blue colour around the edge of the butterfly's wings?
- If yes, it is the Small Tortoiseshell (**E**).
- If no, it is the Peacock butterfly (**D**).

Top Tip!

Learn about how keys work so that you can apply the rules to anything – and don't be put off if the plants or animals you are asked to identify are unfamiliar.

level 5

Branching diagrams

- The other sort of key commonly used is a branching diagram which identifies species from their **characteristics**.
- In this example, a branching diagram is used to identify some types of wild flowers.

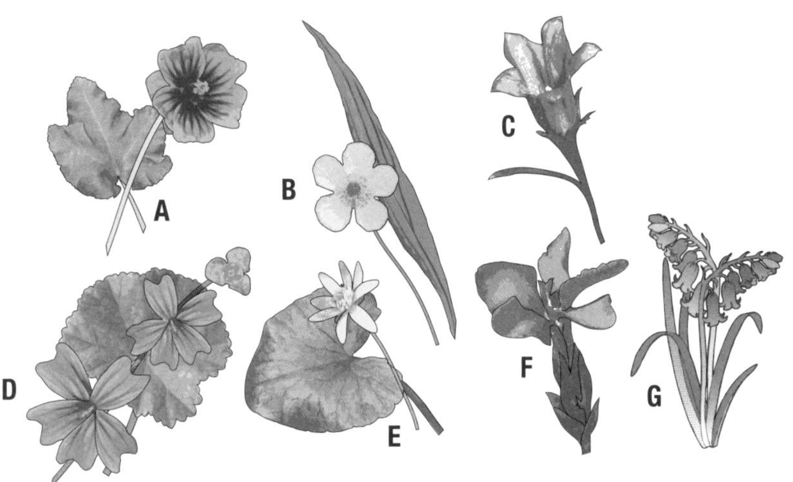

Did You Know?

The venus fly trap is a plant that traps and digests insects.

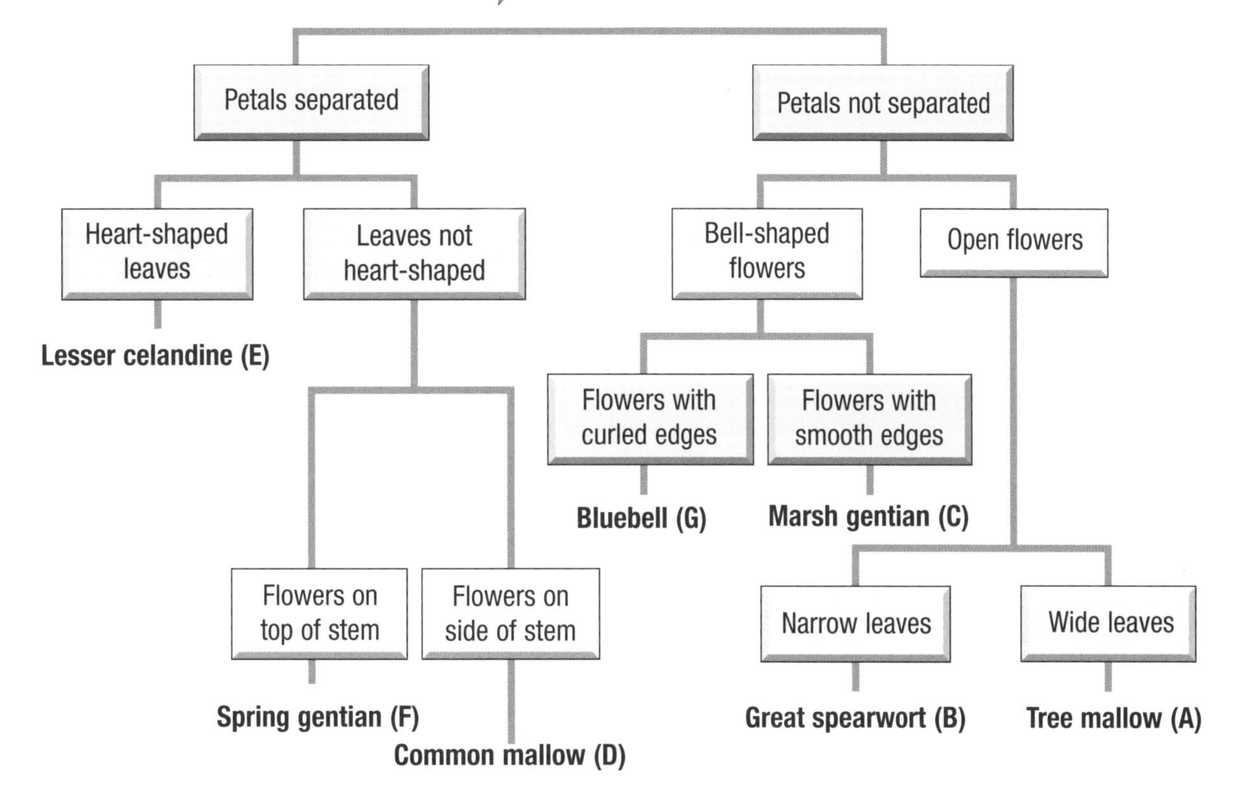

Use the keys on these pages to answer these questions.

1. Which butterfly has blue edges to its wings?
2. Which plant has separate petals and heart-shaped leaves?
3. Which plant has open flowers and wide leaves?

Competition among living things

level 4

Consequences of competition

- Over the centuries, many species of animals and plants have become **extinct**. This means that there are none of these animals or plants left. Many other species are **endangered** which means that there are so few individuals left that the species may become extinct.
- Sometimes the reduction in numbers is due to human impact on the environment (see page 82) and sometimes it is due to other species being better adapted to live in that habitat.

level 4

Woodlands

- In woodland, the trees will often block out the light and prevent plants nearer the ground from being able to make their own food by photosynthesis.
- In a tree, there is only room for a limited number of nest sites for birds, which means that birds have to compete for space as well as food. If there is not enough food, then the number of individuals will be reduced as some will starve.

The leaves of woodland trees block the light so fewer plants grow well under them.

Top Tip!

This is another topic where you can be asked questions about all sorts of situations – learn the basic facts well and if the question is about something unusual, read it very carefully as there is sure to be information to help with the answer.

level 5

Interdependence

- If the numbers, or **population**, of one species goes down, then it has an effect on other creatures that are part of the same food web. Think about the food web in a woodland habitat (see page 69), where small birds and mammals feed on nuts and insects, and large birds and mammals feed on the small ones. Think about the effect of a reduction in the population of greenfly. If there were fewer greenfly, then all of these things might happen:
  - The blue tits would eat more hazelnuts so there would be less for the squirrels and mice to eat.
  - This would in turn mean more competition so the population of all three species might go down.
  - There would be less leaf matter eaten by the greenfly so the mice who can eat leaves as well would be in a better position.
  - If there were fewer blue tits then the owl population might go down – or the owls might eat more mice and then there would be less for the foxes to eat.

level 5

Changes in population

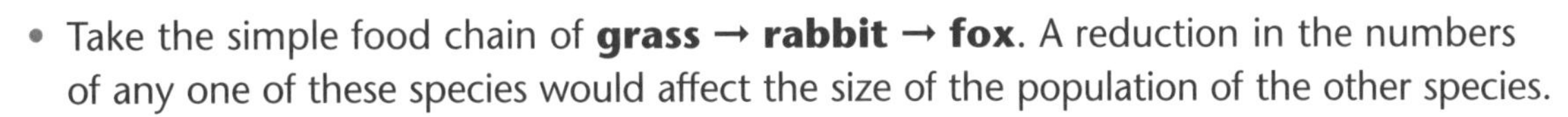

- Take the simple food chain of **grass → rabbit → fox**. A reduction in the numbers of any one of these species would affect the size of the population of the other species.

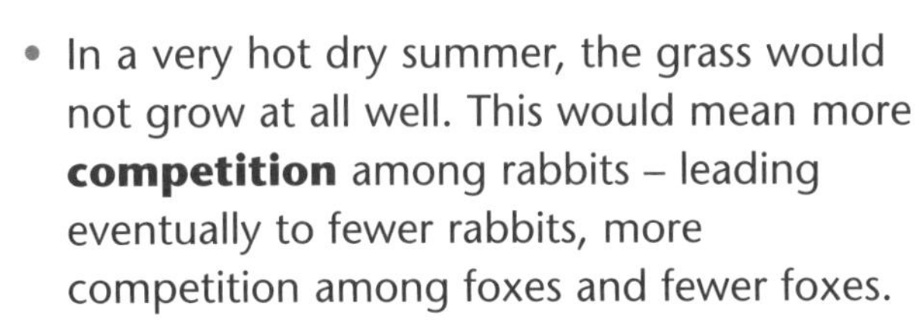

- In a very hot dry summer, the grass would not grow at all well. This would mean more **competition** among rabbits – leading eventually to fewer rabbits, more competition among foxes and fewer foxes.

  A graph of the rabbit and fox populations might look like this:

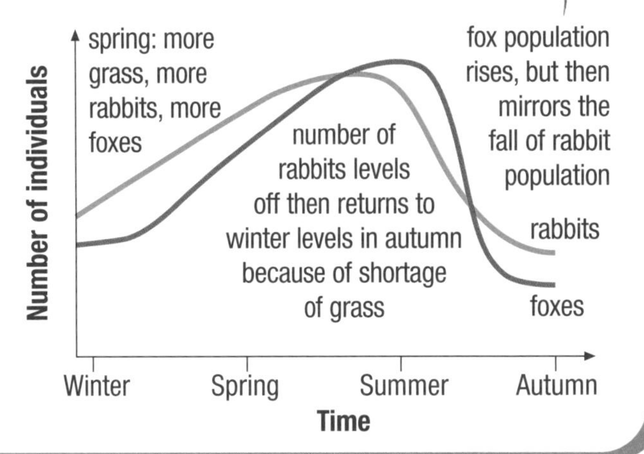

level 6

Other factors

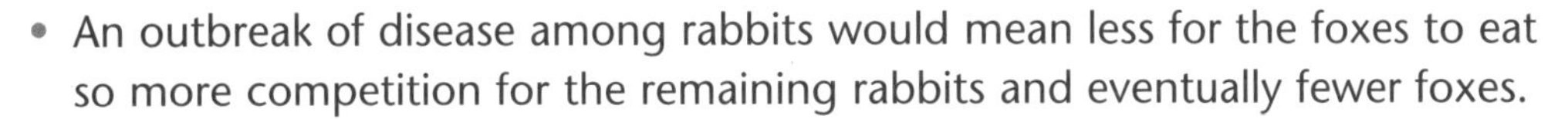

- An outbreak of disease among rabbits would mean less for the foxes to eat so more competition for the remaining rabbits and eventually fewer foxes.

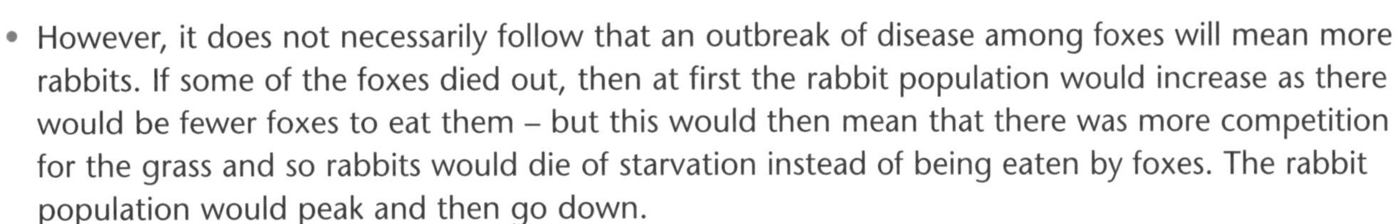

- However, it does not necessarily follow that an outbreak of disease among foxes will mean more rabbits. If some of the foxes died out, then at first the rabbit population would increase as there would be fewer foxes to eat them – but this would then mean that there was more competition for the grass and so rabbits would die of starvation instead of being eaten by foxes. The rabbit population would peak and then go down.

  The graph of the two populations might look like this:

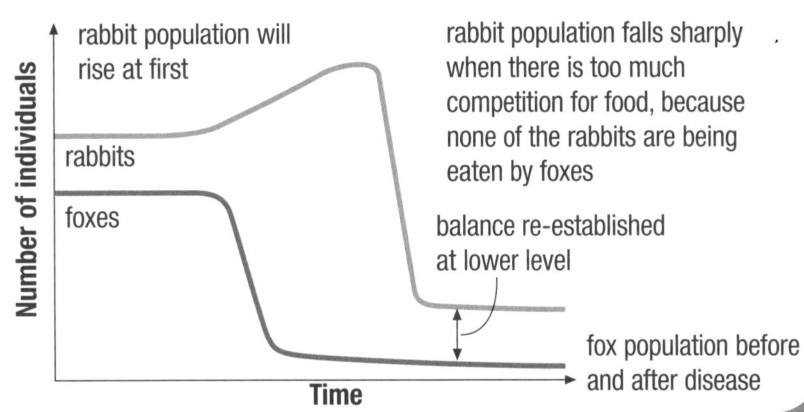

Did You Know?

Another two species of animal become extinct in Britain alone each year.

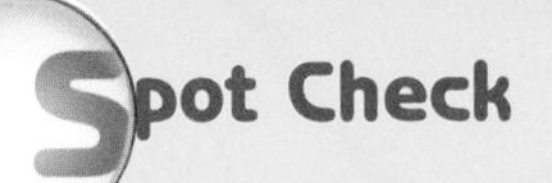

Spot Check

1. In a woodland habitat, a very wet summer may mean fewer hazelnuts on the trees. What affect might this have on the fox population?
2. Plants like bluebells and primroses grow in woodland conditions. Why do they flower in early spring?
3. What do we mean if we say a species of plant or animal is endangered?

Human impact on the environment

level 5

Competition from humans

All the animals and plants that have become extinct have suffered too much **competition** from another species. One of the species that is most in competition with other living things is us – humans.

level 5

Competition for space

- As the human population increases, the effect we have on the environment increases too.
- We compete for space in some parts of the world. People cut down forests and woodlands to use the land to grow crops or for grazing. Competition for space can also take the form of overgrazing when too many animals have to graze on too small a piece of land.

level 5

Climate change

- We still get most of the energy we need to heat our homes and offices – and for transport – from burning fossil fuels. Burning these fossil fuels produces carbon dioxide. Cutting down forests reduces the trees available to take in carbon dioxide for photosynthesis.
- All of this leads to more carbon dioxide in the upper atmosphere. The carbon dioxide acts like a blanket insulating the Earth, which leads to something called the **greenhouse effect**. So an increase of carbon dioxide could lead to the polar ice caps melting, as well as to changes in rainfall and wind patterns.

The greenhouse effect

level 5

Acid rain

- Some of the other waste gases from burning fossil fuels dissolve in the water in the atmosphere and create **acid rain**. When this acid rain falls on lakes and rivers it makes the water more acidic.
- Many species of fish and plankton are not adapted to cope with acidic conditions. Unless something is done to **neutralise** the water, they die. In this picture, limestone is being sprayed onto a river in Sweden to neutralise the acidity in the water.

level 6

Pesticides

- **Pesticides** are chemicals used to kill insects and pests (e.g. slugs) in the garden and on farms. They do not easily break down and so they are passed up the food chain. At each level in the food chain, the organisms eat more than one of those in the level below, so the amount of pesticide builds up. This means that the pesticide kills creatures other than the ones it is targeted at, for example the birds that eat the pests and insects.

Did You Know?

There are about 4000 species of fish living on the Great Barrier Reef off the coast of Australia.

level 6

The ozone layer

- High up in the atmosphere, a layer of a gas called **ozone** protects the Earth from damaging radiation from the Sun. A group of chemicals called CFCs, which used to be used in aerosols and fridges, can damage this protective layer and let the dangerous radiation through. This can cause skin cancers in humans as well as harming other species.

Top Tip!

Don't confuse the greenhouse effect and global warming with damage to the ozone layer. Check you know the causes and effects of each.

Spot Check

1. Which gas is responsible for the greenhouse effect and what is the biggest source of this gas?
2. A pesticide was sprayed onto a crop in a field. A few weeks later it was noted that a large number of sparrowhawks in a nearby wood were dying. The sparrowhawk eats only mice and other small mammals. Could the deaths of the sparrowhawks be caused by the pesticide?
3. What is the name of the gas that makes up the protective layer in the upper atmosphere?

Asking questions and making predictions

level 5

Asking clear questions

- Before you start on any **experiment** it is important to be very clear what the question is that you are **investigating**.
- For example, it is not enough to say that we want to investigate the effect of sunlight on plants; or of acids on alkalis; or of forces on motion.
- The question must explain exactly what you want to find out, what you are going to change and what you are going to measure.
- For example, you could ask:

  A Does the amount of sunlight that a plant receives affect the rate of photosynthesis, as shown by the amount of growth?

  B What quantity of different alkalis will be needed to neutralise 100 ml of hydrochloric acid?

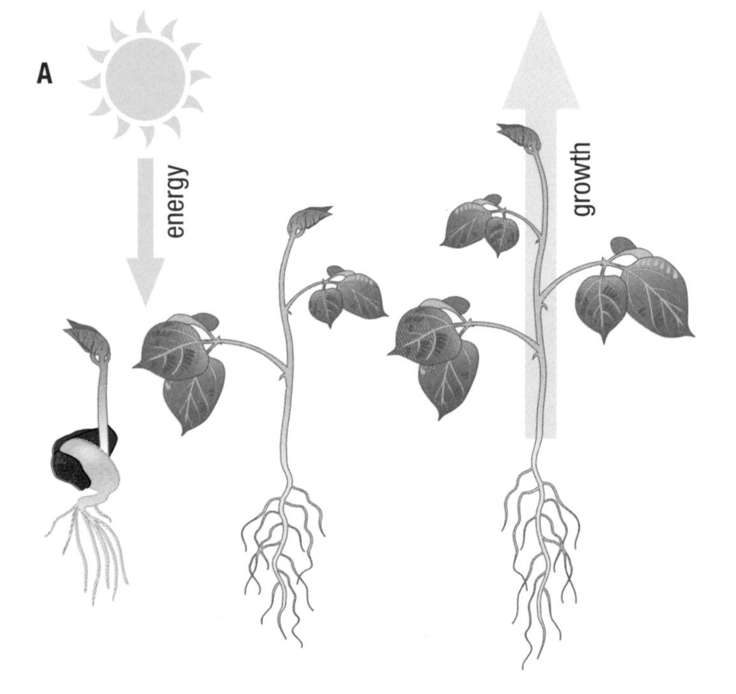

B

alkali

alkali

hydrochloric acid

- The science you learn as theory has all been confirmed by someone doing an experiment. That is what science is. This means that you must keep all the science you know in your head while planning and carrying out experiments.

Top Tip!

In any experiment the factor we choose to change is called the independent variable, the factor we choose to observe or measure is called the dependent variable and the factors we keep the same are called control variables.

Did You Know?

The word 'hypothesis' comes from two Greek words – 'hypo' meaning under and 'thesis' meaning study. So a hypothesis is an idea 'under study' or one being tested by experiment and observation.

level 5

Making predictions

When you have decided on a clear question, it becomes much easier to make a **prediction**. Another word for prediction is **hypothesis**.

- Your prediction must explain what you think will happen and *why* you think it will happen.
- The 'why' must be based on good science. Even if the prediction turns out not to be correct when you do the experiment, this is less important than the quality of your prediction.
- Predictions for the examples on the opposite page could be:

A If the plant gets more hours of sunlight, it will be able to make more food and grow more. This is because light is essential to the chemical reaction of photosynthesis which is how plants make food for growth.

B The stronger the alkali, the less of it will be needed to neutralise the acid. When you mix substances with different pH values, the resulting solution has a pH value that lies between the two original values. An alkali with a high pH (12) will have more neutralising effect than the same quantity of an alkali with a lower pH (8). A neutral solution has a pH of 7.

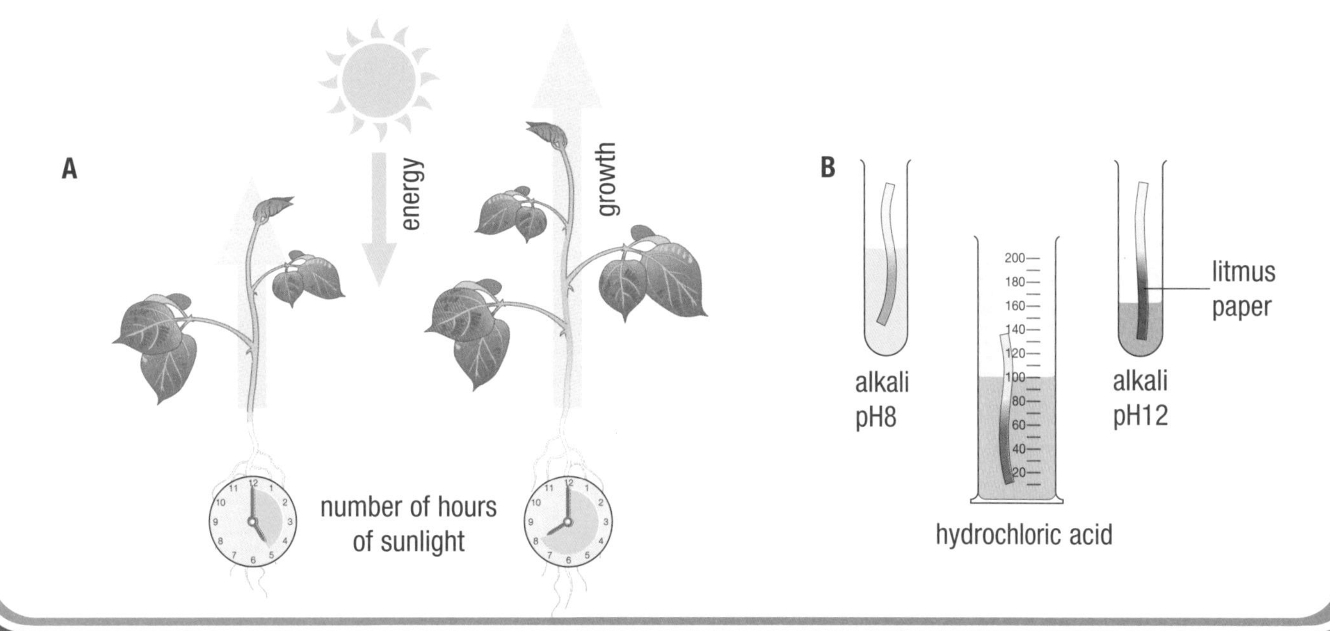

Spot Check

1. Write a suitable question for an enquiry into the effect of temperature on the solubility of salt.
2. Write a hypothesis or prediction for this question.
3. Some students wanted to know which kitchen towel was the most absorbent. Their question was – Are blue paper towels more absorbent than white ones? Can you help them to write a better question?

Interpreting results and reading graphs

level 4

Reading bar charts and pie charts

- Being able to plot the correct graph or choose the correct type of chart to use is very important. But it is also essential to be able to read or **interpret** graphs and charts as this tells us what has happened in the experiment.

level 4

Pie charts

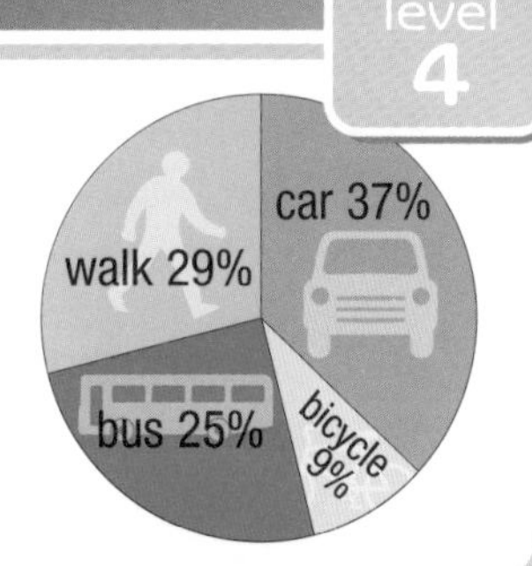

- Pie charts are easy to read – the bigger the slice of pie, the more there is of that group or quantity.
- In a travel survey carried out by a school, a lot of the students travel by car and many walk, nearly as many come by bus, but very few cycle.

level 4

Bar charts

- Similarly, in this bar chart of the temperature rise in a test tube of water caused by burning a variety of foods, the taller the bar, the more energy there is.
- The water is heated most by the cheese and crisps, a bit less by the bread and a lot less by the crispbread – so we can see which food released the most energy when it was burned.

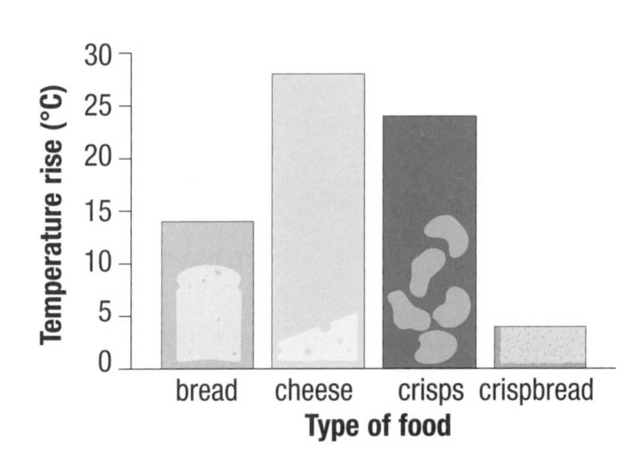

Did You Know?

Edward Jenner (1749–1823) noticed that people who caught cowpox didn't get smallpox. From this observation, he developed the idea of using cowpox as a vaccination against smallpox. Vaccinated people developed a mild form of cowpox, which stopped them getting the more serious smallpox. Jenner's idea is used for present-day vaccinations against many diseases.

Top Tip!

If your independent variable is discontinuous, then a bar chart is probably fine. Continuous variables need line graphs – with a line joining the points (see page 88 for lines of best fit).

level 5

Interpreting line graphs

- Pie charts and bar charts give a picture – but line graphs can tell a story over time and can show how one thing can change in response to the change of another.
- In the graph of temperature and time, we see how the temperature of the water changes over time, levelling off when the ice was melting and again when the water was boiling.

level 5

Giving a scientific explanation

- You always need to give a scientific explanation for what you discover. For the temperature and time graph, the temperature levels off because changes of state use up most of the energy being supplied, so little is left to raise the temperature at these points.
- In the same way, a graph of speed against time can tell us the story of a journey.
- The person doing this journey starts by accelerating until a speed is reached at B, which stays steady until C. She then slows down to a new steady speed which is maintained from D to E. At E she starts to accelerate again and reaches the highest speed of the journey at F. She holds this speed until G when she slows down very quickly to stop at H. She remains still until I, accelerates to J, then slows down again to K, where she stops at her destination.

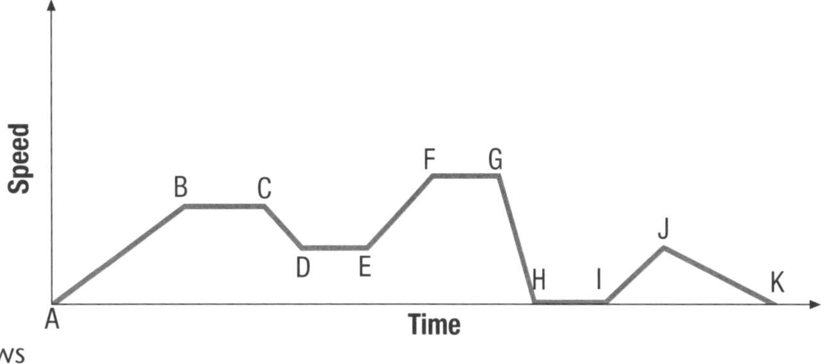

Spot Check

Look at the graph and answer the **three** questions below.
The graph shows how much a spring extends when masses are added to it.

1 What variable should be on the *x*-axis?
2 What should be on the *y*-axis?
3 What do you think has happened at the point where the graph stops going up and becomes a flat straight line?

Evidence and conclusions

level 5

Conclusions

- The results of an experiment can help us to draw **conclusions** about the effect of one variable on another.
- When the results of an experiment show a clear trend or pattern, we can express this as a conclusion. We can conclude from the temperature and time graph on page 87 that if you heat ice, the temperature of the water rises most of the time when it is above 0 °C and below 100 °C, but at 0 °C and 100 °C the temperature levels off.
- The explanation for this is that the changes of state use up most of the energy being supplied so there is little energy left to raise the temperature at these points.

level 5

Lines of best fit

- When our results give a graph that is not a smooth curve or a straight line, any trend will not be as clear. For these graphs, we often draw a **line of best fit** – the line follows the trend of the graph without joining up every point.

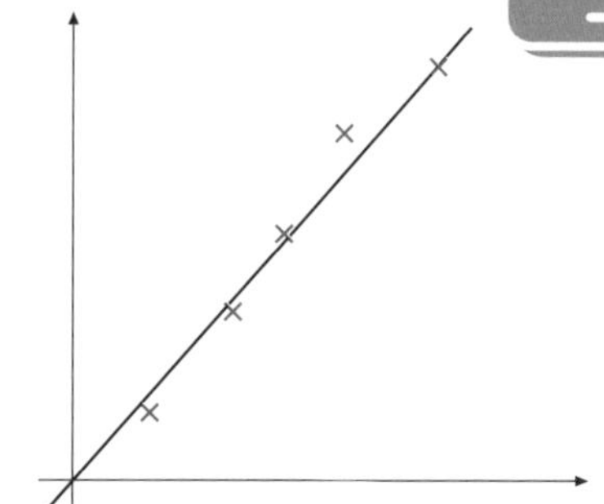

level 5

Evidence

- Results from experiments are examples of **evidence**.
- It is important when drawing conclusions to only use the evidence that is there – not to draw on other things that we might know but which are NOT shown by the results.
- Evidence leads to the development of scientific theories, but the conclusions from one experiment would not be enough for the development of a new theory. It may support a theory or show an area where the theory was weak and more evidence needs to be gathered.

level 5

Opinions and evidence

- It is important to understand the difference between what someone thinks (their opinion) and evidence. To say that smoking is bad for you is an opinion. To be able to show that the number of smokers developing lung cancer is much higher than the number of non-smokers developing lung cancer is evidence to support that view.

Did You Know?

Sir Isaac Newton described how his work built on that of earlier scientists saying: 'If I have seen further than other men it is because I have stood on the shoulders of giants'.

level 6

Development of scientific ideas

- Scientific **ideas** and **theories** are NOT completely fixed and unchanging.
- All the scientific ideas that we have today have been developed over time – sometimes over hundreds of years – as scientists have done more experiments and thought more about the results of experiments done in the past.

level 6

Developing the laws of motion and gravity

- Look back to page 49 where we examined what affects the speed of falling objects.
- A man called Aristotle, who lived more than 2000 years ago in Greece, decided that heavier objects would fall faster. But about 500 years ago, in Italy, Galileo Galilei did some experiments, dropping light and heavy objects from a high building. He concluded that the mass of an object did not affect the speed at which it fell to the ground. His evidence changed people's ideas away from what Aristotle had thought before.
- Sir Isaac Newton, who lived just after Galileo, did more experiments and more work, and developed the laws of motion and gravity that we use today. His work built on what Galileo had done before.
- This is how scientific ideas and theories develop. Scientists challenge or develop the work that others have done before them.

Galileo Galilei

Sir Isaac Newton

Remember that scientific ideas develop over time. What we know now is based on work from the past – and the work being done now will shape the ideas of the future.

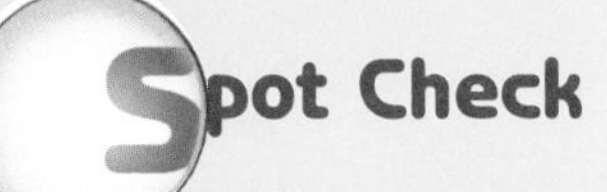

1. In a survey of 25 people, none of the males had naturally blond hair. Is this evidence to support the statement that no men have blond hair?
2. The number of children catching measles each year throughout Great Britain is recorded. If this number went down a great deal in the year that a vaccination was introduced and given to all children, and stayed down – would this be evidence that the vaccination was effective?
3. Where was Galileo's hypothesis finally proved right?

Evaluating experiments

level 6

Reliability and evaluation

- If experiments can be used as evidence to change scientific ideas, then it is important that the conclusions are **reliable**.
- Working out how reliable the conclusions are from an experiment is called **evaluation**.
- Evaluation involves looking at how good an experiment has been – there is no such thing as a perfect experiment. It sometimes gives ideas as to what further work could be done to check and develop those conclusions.
- When you are evaluating your own work – or in a Test evaluating an experiment that has been described to you – there are some things you need to ask yourself.

level 5

Was the test really a fair test?

- Was it possible to control all the control variables completely?
  - Sometimes it isn't possible to do this.
  - For example, in any experiment that involves people, or most other living things, it can be very hard to control all the variables.
  - When an experiment involves working in the dark – how dark is dark?

Top Tip!

Experiments can always be improved. It doesn't mean you got it wrong if you can see faults in your experiment. It means you are well on the way to becoming a real scientist.

level 5

How easy was it to take accurate measurements?

- Was the thing that you were trying to measure easy or difficult to work on?
  - Measuring something that stays still is easy, but when something is moving it is much harder.
  - Measuring a regular shape is much easier than measuring an irregular one.

level 5

How easy was it to take accurate readings?

- Was it possible to take readings that were **accurate**?
  - Did your measuring instrument enable you to take readings that showed up differences in your dependent variable?
  - If you were measuring very small weights with a Newton meter accurate to only 1 N, or small changes in temperature with a thermometer that only measured to 1 °C, would you be able to notice small differences?

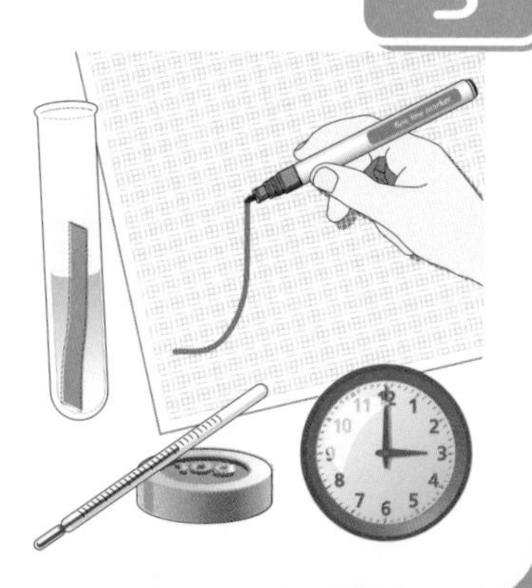

Did You Know?

Some experiments have to be done deep underground at the bottom of disused mines so that nothing on the surface of Earth can interfere with the results.

level 5

Did you take enough readings?

- If you only took two or three readings, then can you really see a trend or pattern or plot a meaningful graph?
- If you took repeat readings, were they close together and, if not, can you see why not? This will tell you about the reliability of your results.

level 5

Did you ask the right question in the first place?

- Was your question appropriate, taking into account the variables that you identified and measured?
- The best evaluations will often lead to further planning and more investigation – and more questions that need to be answered.

level 6

Evaluating real-life experiments

- Evaluation is important in large-scale projects, as well as those that take place in a laboratory.
- A farmer who grows a genetically modified crop will need to ask some questions when the crop has been harvested. Those questions might include:
  - Was the yield (the amount of produce) per acre more or less than in a similar field with a non-GM crop, or than in the same field last year?
  - How much did the crop cost in terms of fertilisers and pesticides and, again, how did this compare with the non-GM crop?
  - Was this crop in any way easier or harder to manage than a non-GM crop?
- All of these answers will include the effect of factors that are not possible to control, such as the weather.

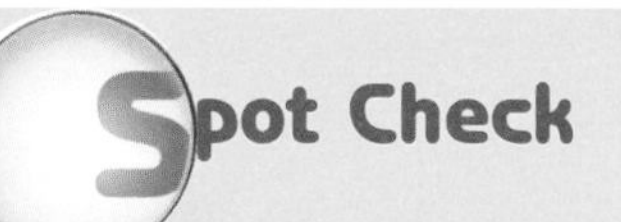

Spot Check

1. In what sorts of experiments can it be difficult to make sure you have controlled all the variables?
2. How can you make sure that experimental readings are as accurate as possible?
3. How can you check to see if your results are reliable?

Answers to Spot Check questions

CELLS

Cell structure pages 4–5

1 Cell membrane; nucleus; cytoplasm.
2 Cell wall.
3 Cell membrane.

Specialist cells pages 6–7

1 They allow the egg cell to move from the ovary to the uterus.
2 Because when they join together in fertilisation, the new cell has the full number of chromosomes, half from each parent.
3 Their bi-concave shape and lack of nucleus means that they have the biggest surface area possible to trap oxygen.

Tissues and organs pages 8–9

1 The nucleus.
2 The brain, the nerves and the sensing organs such as eyes, ears and skin.
3 Kidneys, lungs and skin.

The digestive system pages 10–11

1 It makes a fluid called bile which helps in the digestion of fats.
2 The large intestine.
3 Gastric.

The reproductive system pages 12–13

1 40 weeks.
2 Because the harmful substances in the tobacco can be passed from the mother's bloodstream to the foetus through the placenta.
3 Ovulation, an egg cell is released from the ovary.

The circulation system pages 14–15

1 From the right-hand side, and so the red blood cells can collect oxygen.
2 Capillary walls are only one cell thick as this allows oxygen, carbon dioxide and food and waste products to pass through.
3 Because it reduces the amount of oxygen that the blood can carry and so reduces the amount of energy that can be released from glucose by respiration.

Photosynthesis pages 16–17

1 The leaves are partly green and partly white and only carry out photosynthesis in the green part of the leaves.
2

$$\text{Water + carbon dioxide} \xrightarrow[\text{Chlorophyll}]{\text{Light energy}} \text{glucose + oxygen}$$

3 Because it traps the energy from the Sun and it takes in carbon dioxide which is a waste product of respiration and burning and gives out oxygen needed for respiration and burning.

Photosynthesis and respiration as chemical reactions pages 18–19

1 Oxygen.
2 At night when there is no light for the plant to carry out photosynthesis it will still be respiring so it will be taking in oxygen and giving out carbon dioxide, and this was thought to be unhealthy.

ENERGY

Energy resources pages 20–21

1 Advantages: either concentrated sources of energy OR that they are very flexible and can be used anywhere.
Disadvantages: either that they will one day run out OR they produce gases which harm the environment.
2 Production of CO_2.
3 The greenhouse effect.

Generating and using electricity pages 22–23

1 The steam in a power station turns the turbine which then turns the generator which makes the electricity.
2 There is no combustion (burning); the turbine is turned directly.
3 Loft insulation/double glazing/shutting doors and windows/cavity wall insulation/draft excluders.

Electrical circuits pages 24–25

1 A parallel circuit, as if one bulb broke the others would all stay alight.
2 The one that had just been switched off would still be warm.
3 Chemical energy.

Energy transformation pages 26–27

1 Stored energies are called potential energy.
2 Heat or thermal energy.
3 The transfer of light energy does not need particles but the transfer of sound energy does.

Energy transfer pages 28–29

1 Heat or thermal energy.
2 The heated particles get more energy and so move more quickly and collide with other particles and so pass the energy on.
3 Convection.

Energy from food pages 30–31

1 Because only green plants can trap the energy of sunlight.
2 By burning the same amount of a number of different foods and using the energy to heat the same amount of water each time and measuring the increase in the temperature of the water.
3 Fibre.

Sound and light pages 32–33

1 Same loudness, different pitch.
2 Because we see things when light falls on them and is reflected into our eyes – if there is no light source in the room there will be no light to be reflected.
3 Because light travels faster than sound.

Reflection, refraction and seeing colour pages 34–35

1 Violet.
2 Red.
3 An object looks white because it reflects all the colours of light. When there is only red light shining on it, it will reflect the red light and so appear red.

FORCES

Balanced and unbalanced forces pages 36–37

1 Speed, direction and shape.
2 It will continue to move at the same speed in the same direction for as long as the forces are balanced.
3 By making the surface area smaller.

Friction pages 38–39

1 All except **b**.
2 The push force will be greater than the friction force when the block starts to move.

Streamlining and air resistance pages 40–41

1 Air resistance.
2 Terminal velocity.
3 Weight (or gravity).

Speed and motion pages 42–43

1 Speed = distance ÷ time
= 150 miles ÷ 3 hours
= 50 mph (don't forget the units)
2 No this is the average speed; the speed of the train will have been different at different parts of the journey. Sometimes it will have been going much faster than 50 mph and at other times much more slowly.
3 Metres per second (m/s).

Speed and stopping distance pages 44–45

1 Thinking distance and braking distance.
2 If the driver is tired/has been drinking or using drugs/is using a mobile phone.
3 If the road is wet then there will be less friction between the tyres and the road. This will mean that the car takes longer to stop and in that time travels further.

Pressure and moments pages 46–47

1 P = F/A
= 200/10
= 20 N/cm^2
2 Pressure from boot = F/A
= 200/350
= 0.6 N/cm^2
So spade has increased pressure by 20 – 0.6 = 19.4 N/cm^2.
3 Size of force and distance from the pivot.

Gravity pages 48–49

1 **a** 700 N; **b** 117 N.
2 The shape and size of the two objects so that one did not have more air resistance than the other.

The Earth and beyond pages 50–51

1 Mercury and Venus because they are closer to the Sun so it takes less time for them to complete an orbit so the year is shorter.
2 Because as telescopes have got better we have been able to make better observations.
3 Scientists look for new evidence all the time. Sometimes a new piece of evidence is found that means that an existing idea is not correct and it has to be replaced with a new one.

Magnetic forces pages 52–53

1 The size of the current and the number of turns on the coil.
2 You can turn the magnetism on and off.
3 Magnetic North (remember this is not quite the same as the North pole).

PARTICLES

Solids, liquids and gases pages 54–55

1 **a** Melting; **b** condensing.
2 It is a fixed regular lattice with particles only able to vibrate around fixed points.
3 The particles in a liquid are more free to move than in a solid but not so spread out as in a gas. They can slide over each other and move quite freely but still remain in contact with each other.

Solubility and separation pages 56–57

1 To make a **solution**, take some **solute** and dissolve it in a **solvent**. If your solute will not **dissolve** in your solvent then it is **insoluble**. All solutes will not dissolve in all **solvents**.
2 First filter the solution to get the sand and then evaporate off the water to leave the salt.

Elements, compounds and mixtures pages 58–59

1 On the left-hand side.
2 Carbon, hydrogen and oxygen.
3 One atom of oxygen and one of magnesium.

Chemical reactions pages 60–61

1 Exothermic.
2 Magnesium + oxygen → magnesium oxide

$$2Mg + O_2 \rightarrow 2MgO$$

3 Combustion.

Acids and alkalis pages 62–63

1 You would need to add an alkali. The Indicator is red because the solution is acidic and to turn it green the solution would have to become neutral. The only way to neutralise an acid is to add an alkali.
2 **a** Acid + alkali → salt + water
b Hydrochloric acid + sodium hydroxide → sodium chloride + water
c Acid + metal → salt + hydrogen
d Zinc + sulphuric acid → zinc sulphate + hydrogen

Metals and non-metals pages 64–65

1 Water and oxygen.
2 Metal + steam → metal oxide + hydrogen
3 Magnesium and iron.

Rocks and the rock cycle pages 66–67

1 If the crystals are large then the rock has been formed by slow cooling magma. If they are small, the magma has cooled quickly.
2 Fossils are found in sedimentary rock as the remains of animals and plants get caught up in the pieces of rock that go to form the sediment. As this is turned into rock by pressure over hundreds of years these remains become fossils embedded in the rock.
3 Heat and pressure.

INTERDEPENDENCE

Food chains and food webs pages 68–69

1 A producer is always a green plant – there is always one at the start of a food chain because only green plants can make their own food using the energy from sunlight.
2

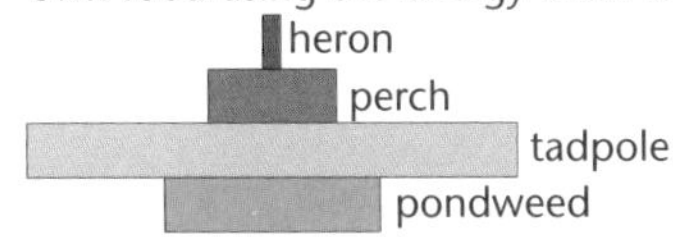

3 The secondary consumers are probably carnivores.

Energy transfers in food chains and food webs pages 70–71

1 An animal that eats another animal is a predator and the animal it eats is its prey.
2 Because at each level some energy is used by the organism for other life processes, e.g. in animals to maintain body temperature and for movement.
3 About 5%.

Variation and inheritance pages 72–73

1 Eye colour and natural hair colour are completely inherited characteristics, height is partly so, as although we inherit a maximum possible height we will only reach that in the right conditions; weight depends on what we eat and how much exercise we take; and anyone can choose to learn a foreign language – these are not inherited.
2 They would not be able to produce any offspring themselves.
3 Quantity and quality of wool produced, hardiness (ability to live outside in winter), quality of the meat produced, etc.

Adaptation pages 74–75

1 Thick coat with waterproof fur/a greenish-brown colour for camouflage/a herbivore/powerful legs for running away from predators, etc., etc.
2 Because there is little food; they eat a lot in the autumn when there is plenty of food.
3 Because in the daytime there will be insects about who will pollinate the plant. At night by closing the flower the plant is protected from the cold.

Classification pages 76–77

1 The young develop inside the mother and are fed on milk from the mother after birth.
2 Insects, arachnids, crustaceans, myriapods.
3 Any two from: warm-blooded, breathe through lungs, feathers, lay eggs with hard shells.

Using keys to identify living things pages 78–79

1 Small tortoiseshell.
2 Lesser celandine.
3 Tree mallow.

Competition among living things pages 80–81

1 If there were fewer hazelnuts then there would be less for the squirrels and mice to eat which might mean less food for the foxes so fewer foxes.
2 Bluebells and primroses flower in early spring before the leaves are all out on the trees so there is more light at ground level than there would be later on when the trees are fully out.
3 There are so few individuals of that species left that it may soon become extinct.

Human impact on the environment pages 82–83

1 Carbon dioxide which is released by burning fossil fuels.
2 Yes they could. The small mammals that the sparrowhawks feed on may well eat the crop growing in the field. The amount of pesticide eaten by each mouse would not be very great and possibly would not harm the mouse. But each time a sparrowhawk eats a mouse, the amount of pesticide it has eaten will increase until after a few weeks the level of pesticide in its body is enough to kill the bird.
3 Ozone.

SCIENTIFIC ENQUIRY

Asking questions and making predictions pages 84–85

1 Does the temperature of water affect the time taken for a given amount of salt to dissolve?
2 If the water is at a higher temperature the same amount of salt will take less time to dissolve as the particles will all have more energy and be able to move about more.
3 What volume of water can be absorbed by a 30 cm square (or any size as long as it's the same for each) of different kinds of kitchen towel?

Interpreting results and reading graphs pages 86–87

1 Mass added to the spring in grams.
2 The length of the spring or the amount of extension in mm.
3 The spring has stretched so far that it is no longer a spring – adding more and more mass will not make it any longer.

Evidence and conclusions pages 88–89

1 No, because not enough people were asked.
2 Yes, because the sample size is very large and there has been a clear change – the introduction of the vaccine – and a clear effect – the reduction in the number of cases of measles.
3 From the top of a tall building.

Evaluating experiments pages 90–91

1 Experiments involving living things.
2 By making sure that you use the correct measuring instrument.
3 By repeating them to see if you get the same result each time or if they are quite different.

Glossary

accuracy the quality of the measurements taken in an experiment

acid rain rain containing dissolved gases such as sulphur dioxide, which are produced when fossil fuels are burnt

air resistance the force that slows down objects moving through the air

ammeter a device for measuring electrical current

amplitude the height of a sound wave

arteries thick-walled blood vessels that carry blood away from the heart

average speed the total distance travelled divided by the total time

balanced forces when the forces acting on an object do not cause it to change its speed or direction

braking distance the distance a car travels before it comes to a halt after the brake has been applied

breathing taking in air, rich in oxygen, from the surroundings and expelling air, rich in carbon dioxide, from the body

capillaries tiny blood vessels with walls one cell thick

carbohydrate the food needed for energy

carnivore an animal that eats only other animals

cell membrane found in plant and animal cells and controls what goes into and out of the cell

cell wall rigid structure made of cellulose outside the cell membrane that gives plant cells support

chloroplast contains the green pigment chlorophyll found in plant cells that is needed for photosynthesis

chromatography a method of separating different solutes from one solvent where the particles of the solutes are of different sizes

chromosomes structures containing DNA found in the nucleus of cells

combustion the reaction that happens when substances burn in oxygen

competition trying to obtain the resources needed to survive at the same time as others

compound a substance formed when two or more elements react together chemically

conduction thermal energy transfer through solids

conductor thermal conductor – a material through which heat flows easily; electrical conductor – a material through which electrical current flows easily

control variable the quantity in an experiment that is kept the same to make sure that the effect of the independent variable is measured

convection thermal energy transfer through liquids and gases

corrosion the reaction of a metal with oxygen to form an oxide

dependent variable the quantity in an experiment that is measured to see the effect of the independent variable

digestion the process of breaking food down into simple substances that are used for growth and to provide energy

digestive system the organs that work together to break down the food an animal eats

displacement reaction a reaction where a metal that is higher in the reactivity series replaces another that is lower in the reactivity series from a salt

distillation a method of separating liquids which have different boiling points

electromagnet a material that has a magnetic field only when an electric current is flowing through it

element a pure substance made entirely of one type of atom

embryo a ball of cells formed when a sperm fertilises an egg and which becomes implanted in the uterus

evaluation an assessment of how accurate and reliable an experiment has been, and how that experiment could be improved

evidence facts and factors that can be measured to prove or dispute a theory or an idea

filtration a method of separating large particles from small ones – used to separate a solid from a liquid

foetus the unborn baby, older than 8 weeks, developing in the uterus

food chain a series of organisms, always starting with a plant, each one feeding on the next one down the chain to obtain the energy needed for life processes

food web the interaction of several food chains reflecting the variety found in natural situations

formula the name of a compound written using chemical symbols

frequency the number of sound waves passing a particular point in a second

friction the force that acts to oppose movement

global warming the increase in the temperature of the Earth's atmosphere caused by the greenhouse effect

gravity the attraction force between any two objects due to their mass

greenhouse effect the insulation effect of the layer of carbon dioxide in the upper atmosphere

habitat the place where plants and animals live

haemoglobin a chemical found in red blood cells that combines with oxygen to carry it round the body

heart the muscular pump that moves blood around the body

herbivore an animal that eats only green plants

hibernation an inactive state that allows animals to survive in winter

independent variable the quantity in an experiment that is changed to study its effect

joule the unit used for measuring energy

kilocalories the unit used to measure the energy stored in food

kinetic energy the energy of movement

lava molten rock above ground

light intensity a measure of the amount of available light

loudness related to the amplitude of a sound wave

luminous objects that give off their own light

magma molten rock underground

mass how much there is of something, measured in grams

menstrual cycle a 28-day cycle for receiving a fertilised egg in the uterus

menstruation (period) the breakdown and loss of the uterus lining through the vagina

migration the movement of animals to another place where conditions for survival are more favourable

mixture two or more elements or compounds that can be easily separated because they have not undergone any chemical change

molecule two or more atoms joined together

moments a force which acts to make a turn about a pivot

neutralisation the reaction of an acid with an alkali to produce a salt plus water

newton the unit of force – named after Sir Isaac Newton

nocturnal animals that are active at night-time

nucleus the part of a cell that contains the genetic information and controls the chemical activity of the cell

opaque a material that light does not pass through, so it cannot be seen through

organ a group of different tissues joined together to carry out a role within a plant or animal, such as a leaf or root, the heart or stomach

ozone layer a layer of ozone gas high in the atmosphere that protects the Earth from the damaging radiation of the Sun

parallel circuit an electrical circuit in which there are two or more possible routes for the current to flow along

Periodic Table a table of all the elements arranged according to the size of their atoms and their chemical properties

photosynthesis the process by which plants make their own food from carbon dioxide and water by trapping the energy from the Sun

pitch related to the frequency of a sound wave

pivot the point around which turning forces act

potential energy stored energy

predator an animal that eats other animals

prediction a statement about the possible outcome of an experiment based on scientific knowledge

pressure the relationship between the force acting on an object and the surface area over which the force is applied

prey an animal that is eaten by other animals

producer the first organism in a food chain, which is always a green plant

protein the type of food needed for the growth and repair of our bodies

radiation thermal energy transfer from an object that is warmer than its surroundings

reactivity series metals placed in order of how reactive they are with water, steam and acids

reflection the way in which light rays bounce off objects, which enables us to see them

refraction the way in which light rays bend when they move from one medium to another

reliability how easy it would be to obtain the same results if an experiment was repeated

renewable energy sources types of energy resources such as wind and tidal power which will not run out in the way that fossil fuels will

reproduction the process by which living things produce new organisms like themselves

reproductive system the organs used by a plant or an animal to reproduce

respiration the chemical reaction that takes place in every cell to release energy from the products of digested food

rusting the corrosion of iron

salt a chemical formed when an acid reacts with an alkali

series circuit an electrical circuit in which all the components are on one continuous wire

solute a solid material that can be dissolved in a liquid

solution a liquid containing dissolved particles of a solid

solvent a liquid in which a solid material can be dissolved

species animals, or plants, that can reproduce to give offspring that are capable of reproduction

spectrum the seven colours produced when white light is split up using a prism

speed the distance an object travels in a particular time

stopping distance the total distance that a car travels in the time taken between the driver deciding to stop and the car coming to a halt

terminal velocity the maximum speed that any object can reach

thermal energy the proper way of describing heat

thinking distance the distance a car travels in the time between the driver deciding to stop and applying the brake

tissue a group of similar cells in a plant or an animal

translucent a material that light can pass through

trophic level the steps in a food chain

unbalanced forces when a force in one direction is greater than the forces in other directions, which causes the object to change its speed or direction

vacuole bag-like structure in the centre of a plant cell that helps to keep it rigid

veins vessels with valves that carry blood back to the heart

weathering the breakdown of rock by rain and wind

weight a downward force, measured in newtons, caused by gravity acting on an object

Index